Time and the Multiverse

Julian von Abele

Copyright © 2017 Julian von Abele
All rights reserved.
ISBN: 1545464820
ISBN-13: 978-1545464823

DEDICATION

To Leonard Euler, the master of them all.

CONTENTS

1 Preface   Pg #3

2 Generalization of Path-Integration   Pg #9
3 A One-Electron Theory   Pg #35
4 Path-Integral Formulation of Temporal Dynamics   Pg #69
5 Superluminal Travel and Communication   Pg #117

ACKNOWLEDGMENTS

Many thanks to Dr. Michael Carchidi for painstakingly reviewing this work and the ideas therein. I also express gratitude to my parents for offering encouragement during my research projects.

Preface

Mathematics is fundamental to the structure of reality. Mathematical advances, in turn, have proven central to discoveries in theoretical physics. Generalizations of mathematical structures—the extension of Number from the rationals to the reals to the complex, for example—have clearly developed foundational roles in physical theories. While the notion of complex numbers once confounded the mathematicians of the eighteenth century—indeed, leading to their demonym "imaginary"—they today form the edifice of quantum mechanics, and the mathematical framework of quantum theory in general. While the notion of exponentiating a complex number might once have been met with similar dismissal, the generalization of exponentiation to complex power led to Euler's formula, and all the ramifications and applications that discovery has entailed. The historical record is clear: generalizations of mathematical concepts, even apparently bizarre generalizations, lead the way to future discoveries, and these future discoveries often serve as foundations of physical theories.

The present work describes several new hypotheses pertaining to theoretical physics, resulting from generalizations of mathematical frameworks. The first concerns extending the process of integration to a complex number of variables, or a domain of complex dimension, much in accordance with the general programme of the fractional calculus. This mathematical generalization, accomplished through the systematization of properties of multiple integration over an integer number of variables, lent itself naturally to application to the path-integral formulation of quantum mechanics. The latter framework determines the probability amplitude as a functional integral, itself determined through a discretization process. By allowing the discretization index to be complex, and applying the aforementioned calculus, we discovered a spectrum of results for the probability amplitude corresponding to distinct limits of the index in the complex plane. It seems unnatural to restrict this index to the positive integers; much as exponentiation was readily extended to the complex plane, it seems too that the quantum path-integral with complex discretization index should be considered just as important as the real one, in accordance with symmetry.

In the past few decades, a number of theoretical advances have focused on generalizing mathematical frameworks of physical theories, and describing the products of this generalization as corresponding to alternate realities or universes. Steven Weinberg extended the field equations of general relativity to the case of a non-zero cosmological constant, and proposed that each different equation, corresponding to a particular value of this constant, applied in a separate universe. The "landscape" idea in string theory sees each different Calabi-Yau manifold consistent with the mathematical framework likewise corresponding to a different universe. Indeed, the physicist Max Tegmark has suggested that every mathematical structure forms its own universe, with our physical laws constituting only one among a panoply of possibilities.

The mathematical foundations of classical and quantum physics obey the fundamental principle of symmetry. Symmetry diminishes information, rendering physical laws and conditions simpler to explain. If the universe possesses some observable property (for example, the strength of the gravitational force) O, the explanation of O must, by necessity, rest on some complicated mathematical contingency. The fact this universe has property O and not E say represents a violation of symmetry. If, however, this universe is part of a multiverse, the property O is easier to explain. Symmetry is only preserved at a fundamental level if each alternative property O' is also instantiated, which is to say it corresponds to its own universe. Clearly some such properties are results of fundamental mathematical properties (consistency, for example) and there need be no symmetry between themselves and their alternatives O'. However, if each O' is itself equally consistent or motivated by the mathematical framework, symmetry often overrules uniqueness. The symmetry among the instantiation of Calabi-Yau manifolds in the string landscape, or cosmological constant values in Weinberg's multiverse, is not the only example of this phenomenon; the placement of the planets in orbit around the sun is likewise an arbitrary property, whose alternatives are instantiated in other star systems. In this case, symmetry also overrides uniqueness.

If we take R as the property of the quantum structure of our universe, reflecting the real value of the path-integral discretization index, it is not clear why R and not some alternative C is instantiated. The claim that only R is instantiated begs the question of why this particular mathematical structure, among an infinite number of alternatives, is relevant to physical reality. Indeed, the aforementioned calculus allows us to consistently describe the path-integral resulting from a complex discretization index, and hence the alternatives C to R should also be instantiated. This latter contingency implies that each different complex phase of the index applies to a separate universe. The resultant multiverse is the subject of the first part of this text.

The subsequent sections concern themselves with the nature of time. Nonrelativistic quantum mechanics describes space as an observable, but time as a parameter in the Schrodinger equation. In constructing a theory of quantum temporal dynamics, we promote time from a parameter to an operator in its own right, and as a consequence provide a theoretical framework for the one-electron hypothesis of Wheeler. This latter idea arose from the Feynman diagrams of particle interactions, and postulates that all electrons in the universe are actually the same electron, but moving forward and backward in time. We propose a nonrelativistic version of the one-electron universe, by describing the dynamics of a particle's "motion through time" much in the same fashion as quantum theory describes its motion through space. As a consequence, we determine a dichotomy between two separate temporal components, the time external to a system and the time internal to a system. External time indexes the interaction of this single particle with a surrounding system, while internal time marks the evolution of its quantum state.

Indeed, time is, by its very nature, a multi-dimensional entity. Time serves as a coordinate framework for indexing the interactions between particles and systems. We say (in the nonrelativistic universe) that two systems can only interact if they coincide in time. This "external time" is not necessarily equivalent to the "internal time" that evolves the quantum state of a single particle. When a particle moves back and forth in external time, a given observer would perceive many different particles corresponding to the intersection of this world-line with a particular point in external time; indeed, this particle can interact with itself if its past and future instantiations coincide in external time. The single electron reflects between two vast barriers in the distant future and past, producing all the electrons that apparently fill the universe. Beyond these potential barriers might lie other potential barriers, between any two of which corresponds another universe. The resultant cosmological multiverse of quantum temporal dynamics served as a convenient mechanism to introduce the aforementioned multiverse of complex-discretized quantum mechanics, and indeed these two theories are combined in an intriguing fashion in the succeeding section.

In the final part of this book, we describe the bizarre consequences of this theoretical framework. Superluminal travel through the parallel universes of this model appears to be a possibility, and we examine a hypothetical starship that could be used to exploit this property. This should not be taken as a serious proposal, but rather as an amusing consequence of an interesting mathematical framework. The main body of this text is derived from a series of academic papers and articles exploring different facets of these theories. While this is not an experimental endeavour, it is interesting to consider the ramifications of mathematics, and mathematical generalizations, on physical law. This book explores only one small part of this vast frontier.

Generalization of Path-Integration to a Complex Time-Discretization Index

Abstract:

We generalize the path-integration process in quantum mechanics to the case of a complex-valued time-discretization index. As is well known, the propagator for particle movement in the path integral formulation is given as a limit of a multiple integral expression, with a time-slicing index approaching infinity. By allowing for a complex-valued index (a complex *number* of time slices), and augmenting the formula for the action—so as to maintain the status of the path contribution formula as a phase factor—the limit is indeterminate and yields a variety of different results for the propagator. These different propagators are postulated to apply to different universes in an expansive multiverse. This process is completely different from the standard practice of calculating path integrals in imaginary time, as it is the time-slicing index, as opposed to the time interval, that is generalized to the complex numbers; the results (summarized in Fig. 1 and equations (7)-(12)) are novel as well.

I. Introduction and Central Idea

A. Overview

In the path integral formulation of quantum theory, the probability amplitude that a particle will travel from point A to point B over a time interval T is proportional to a functional integral of a contribution formula over all possible paths connecting A and B. Such a functional integral is frequently defined via a time-discretization procedure, whereby the time interval T is divided into an index N number of "slices," and the contribution formula is integrated over all "discrete paths" formed by the slicing process. Such a discrete path is given by straight lines within each slice, connecting the spatial coordinates x_i located at each slice point. Thus, an integral over these discrete paths is given as an integral over the x_i. Allowing the index N to approach infinity effects the functional integral over all possible paths. Therefore, the propagator is proportional to [2]

$$P(a,b) \cong \lim_{n \to \infty} \iiint \ldots \int exp\left(\frac{i}{\hbar}S\right) dx_1 \ldots dx_{n-1} \quad (1)$$

In this case, the contribution formula is given as a phase factor dependent on the action. As Feynman has noted [2], contributions from different paths differ only in phase, not magnitude. The action itself is given as the integral of the Lagrangian over the time-interval T. Indeed,

$$S = \int_0^T L \, dt \quad (2)$$

Here, L is the Lagrangian over time. Often, in physical problems, this integral over time is divided into N different integrals corresponding to each time slice. In general, the Lagrangian L is given as

$$L = \frac{m}{2}v^2 - V \quad (3),$$

where V is the potential, and the velocity v is, between t_{i-1} and t_i,

$$v = \frac{(x_i - x_{i-1})}{T} N \quad (4)$$

The index N is typically taken to be a positive integer. While the study of path integrals in imaginary time (that is, for imaginary time interval T) is well known, the index N describing the time-discretization structure has always been taken to be a natural number. After all, it is difficult to imagine a non-integer "number" of time-slices. The central idea behind this paper is to generalize the index N to the complex numbers. We utilize an analytic continuation to determine the result. In the process, a multiple integral over a complex "number" N of variables is taken, and subsequently a limit as N approaches infinity is introduced. As we shall demonstrate, this limit is indefinite, and has a spectrum of possible results depending upon the way N goes to infinity in the complex plane. These different results are postulated to apply to different universes.

In the case of complex N, the quantity (4), specifying the particle velocity, is complex valued. Therefore, the Lagrangian, as in (3), is also complex, and apparently so is the action (2). Therefore, the contribution formula as it appears in (1) (the integrand), is not only a phase factor, but possesses a real exponential part. This contingency is in violation of Feynman's postulate [2], that the contribution of different paths differ only in phase and not in magnitude. Thus, if we are to generalize (1) to complex-valued N, we must augment the formulas for path contribution (integrand of (1), delineated in (2)-(4)) to maintain the status of the path integrand as a phase factor.

Thus, to keep the action real-valued, we re-define (2) as a magnitude,

$$S = \left| \int_0^T L \, dt \right| \quad (5)$$

The lagrangian L is still given as in (3); however, its complex value (dependent on complex N) no longer disrupts the real nature of the action—thus, under suitable interpretation, the contribution formula remains a phase factor (although there are still a complex number (N) *of variables* in (1), that appear in (5)).

Thus, the contribution formula is given as

$$exp\left(\frac{i}{\hbar}\left|\int_0^T L\,dt\right|\right) \quad (6)$$

thus yielding the standard action in the case of real N[1].

The presence of the magnitude in (6) makes the limit (1), as applied to complex-valued N, indeterminate. Indeed, many complex-valued functions involving a magnitude, for example $z/|z|$, yield indeterminate results as the complex variable approaches infinity. In the case of (1), as we shall show, a spectrum of different results are obtained depending upon how the complex-valued index N approaches infinity in the plane. These different propagators are postulated to apply to different universes.

This generalization may seem arbitrary. However, similar such generalizations have found utility in fundamental physics. For example, an extension of the Einstein field equations to the case of a nonzero cosmological constant led to an anthropic multiverse argument, with different universes corresponding to different values of the constant [7]. Additionally, the well-known landscape concept in superstring theory was a result of postulating the existence of different universes corresponding to different Calabi-Yau manifolds [4], and similar generalizations of physical law have also found support as potentially implying the existence of a multiverse [11]. Mathematically, the use of fractional and complex-order differentiation, roughly analogous to the

---

[1] This formula reduces to the standard contribution formula for N real. As delineated in footnote (3), in the case of potential V, the contribution formula is best defined as a continuation of a power series.

notion of functional integration with a complex discretization index, has found great application in physics [12]. Similarly, this paper sees a generalization of the path-integral time-discretization process to the case of complex valued N. To keep the generalization as natural as possible, we maintain the status of the contribution formula as a pure phase factor. By taking the magnitude of the action, the path integrand remains a phase factor (*under suitable interpretation*—it is not quite clear how to directly quantify the value of the integrand of (1) that may contain a complex number N of variables). The multitude of results depending upon the way by which the index N approaches infinity in the complex plane are interpreted to apply to different universes. This generalization may lead to some advances in the field.

B. Further Generalization of the Contribution Factor

As currently construed, the present construction involves two essential generalizations of the Feynman path integral: firstly, an extension of the time-discretization index N to complex values, and secondly, a corresponding generalization of the path contribution formula, preserving its status as a phase factor. However, the continuation (6) to complex N is only one such possible generalization. Applying the magnitude to the action is not the only process yielding an analytic continuation of the contribution factor. In the complex plane, a variety of magnitudes exist; the Lebesgue p-norm summarizes most useful magnitudes, including the well-known "taxicab" and "maximum" norms. In order to use the most complete generalization, we consider different continuations of the contribution formula, and replace the magnitude in (5) with a general p-norm.

On a complex number $x + iy$, the p-norm is defined as [5]

$$|x+iy|_p = (|x|^p + |y|^p)^{1/p}$$

As such, we adjust (6) to become

$$exp\left(\frac{i}{\hbar}\left|\int_0^T L\, dt\right|_p\right) \quad (7)$$

thus capturing the different analytic continuations that may be applied to the contribution formula, subject to the phase-factor requirement.

II. Free Particle Calculation

We shall utilize this generalization to compute the probability amplitude corresponding to a free particle. Using (1), (7), we obtain

$$K_0 \cong \lim_{n\to\infty} \iiint \cdots \int exp\left(\frac{i}{\hbar}\left|\int_0^T L\, dt\right|_p\right) dx_1 \ldots dx_{n-1}$$

We may divide the integral from 0 to T, that defines the action, into n smaller integrals corresponding to each "time slice." Following through with this procedure, we deduce

$$\lim_{n\to\infty} \iiint \cdots \int exp\left(\frac{i}{\hbar}\frac{m}{2T^2}\left|\sum_{j=0}^{n-1}\int_{t_j}^{t_{j+1}}\{x_{j+1} - x_j\}^2 n^2\, dt\right|_p\right) dx_1 \ldots dx_{n-1}$$

Clearly, assuming that the real or imaginary part of a sum is the sum of the real or imaginary parts of the terms,

$$Re\left(\sum_{j=0}^{n-1}\int_{t_j}^{t_{j+1}}\{x_{j+1} - x_j\}^2 n^2\, dt\right) = Re(n)\left[\sum_{j=0}^{n-1}T\{x_{j+1} - x_j\}^2\right]$$

Likewise,

$$Im\left(\sum_{j=0}^{n-1}\int_{t_j}^{t_{j+1}}\{x_{j+1} - x_j\}^2 n^2\, dt\right) = Im(n)\left[\sum_{j=0}^{n-1}T\{x_{j+1} - x_j\}^2\right]$$

Implying that

$$\left| \sum_{j=0}^{n-1} \int_{t_j}^{t_{j+1}} \{x_{j+1} - x_j\}^2 n^2 \, dt \right|_p = |n|_p \left[\sum_{j=0}^{n-1} T\{x_{j+1} - x_j\}^2 \right]$$

Thus, this integral simplifies to[2]

$$\lim_{n \to \infty} \iiint \ldots \int exp\left(A \sum_{j=0}^{n-1} \{x_{j+1} - x_j\}^2 \right) dx_1 \ldots dx_{n-1}$$

Here, A is given by

$$A = \frac{i}{\hbar} \frac{m}{2} \frac{|n|_p}{T}$$

We now introduce a change of variables, substituting z_j for $x_j - x_{j-1}$. Thus, we have

$$\lim_{n \to \infty} \iiint \ldots \int exp\left(A \left\{ \sum_{j=1}^{n-1} \{z_j\}^2 + \{x_n - x_{n-1}\}^2 \right\} \right) dz_1 \ldots dz_{n-1}$$

Simplifying,

$$\lim_{n \to \infty} \iiint \ldots \int exp\left(A \left\{ \sum_{j=1}^{n-1} \{z_j\}^2 + \left(x_n - x_0 - \sum_{i=1}^{n-1} z_i \right)^2 \right\} \right) dz_1 \ldots dz_{n-1}$$

Use of Fourier integration to decompose the squared exponential produces

---

[2] Here, again, we assume that the real or imaginary part of a sum is equal to the sum of the corresponding real or imaginary parts, even in the case of complex n. We thus generalize the known properties of the real and imaginary parts to the case of a sum with a complex "number" of terms. The resulting sum, although real-valued, depends on a complex number of distinct variables, and thus its value is not well-defined. Similarly, although the Dirac delta "function" is not well-defined over its domain, its integral is; even though it is impossible to quantify the value of a function of a complex number of variables, we may use a separation of variables to assign a value to its integral.

$$\lim_{n\to\infty} \int_{-\infty}^{\infty} exp\left(\frac{k^2}{4A}\right) \iiint \ldots \int exp\left(A\sum_{j=1}^{n-1}\{z_j\}^2\right) exp\left(ik\left(x_n - x_0\right.\right.$$

$$\left.\left. - \sum_{i=1}^{n-1} z_i\right)\right) dz_1 \ldots dz_{n-1} \, dk$$

Separation of variables may now be utilized to compute this multiple integral. Specifically, we calculate

$$\lim_{n\to\infty} \int_{-\infty}^{\infty} exp\left(\frac{k^2}{4A}\right) exp(ik(x_n - x_0)) \left\{\int exp(Az^2 - ikz)dz\right\}^{n-1} dk$$

This yields

$$\lim_{n\to\infty} exp\left(\frac{mX^2 i \, |n|_p}{2\hbar T n}\right)$$

In which X is the total displacement traversed by the particle $(x_n - x_0)$. It is at this point that we ascertain the limit of our complex-valued index n. Specifically, we introduce a substitution of variables, by writing the index in polar form,

$$n = r \, exp(it)$$

Here, r is a real number which approaches infinity as n approaches infinity, while t is a separate real parameter. Thus, we have

$$\lim_{r\to\infty} exp\left(\frac{mX^2 i \, r^2(|exp(it)|_p)}{2\hbar T r^2 exp(it)}\right) = K_{0\,std}^C \quad (8)$$

In which $K_{0\,std}$ is the standard probability amplitude for a free particle, while C is given by

$$\frac{|exp(it)|_p}{exp(it)} \quad (9)$$

C parameterizes a region in the complex plane, located between the parametric curves $\left((|\cos(t)| + |\sin(t)|) \cos(t), -(|\cos(t)| + |\sin(t)|)\sin(t)\right)$ and

$(max(|\cos(t)|, |\sin(t)|)\cos(t), -max(|\cos(t)|, |\sin(t)|)\sin(t)), \pi \leq t \leq 2\pi$, for $p \geq 1$. An image of this figure is provided in **Fig.1**.

Thus, this generalization of the time-discretization index, subject to appropriate definitions of the action in order to yield a phase factor contribution formula, yields a variety of different probability amplitudes, due to the non-uniqueness of the limit. These probability amplitudes are complex powers of the standard kernel, with the exponent defined within a specific region of the complex plane. Alternatively, these particles could be viewed as behaving with a complex mass, with this complex mass being a multiplier of the normal mass, the multiplier appearing within the specific area illustrated in **Fig. 1**.

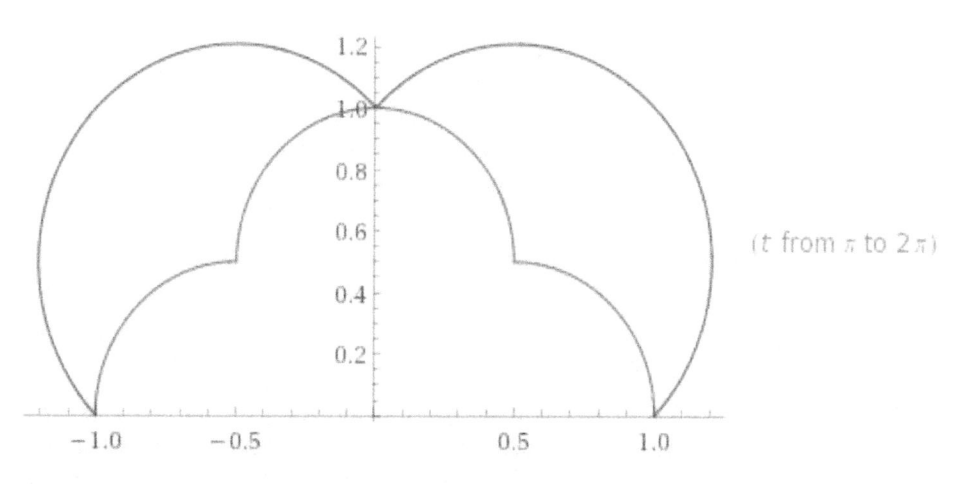

Fig. 1- Contour in the Complex Plane which Delineates Different Probability Amplitudes in Generalization

III. The Particle in a Potential

A. Perturbation Expansion

We shall apply perturbation methods to calculate the path integrals associated with a potential, and analytically extend the results to the case of a complex number of time slices. We begin by noting that the integral of the potential may be expanded as

$$\int V\,dt = \sum_{j=0}^{n-1}\int_{t_j}^{t_{j+1}} V(x)\,dt = \sum_{i=0}^{n-1}\frac{T}{n}\int_0^1 V(x_j + (x_{j+1}-x_j)s)\,ds$$

As n approaches infinity (we will assume a potential only explicitly dependent on position, not time), this expression provides the integral of the potential. Given this expression, we consider the contribution formula (7), and simplify[3]:

$$\exp\left(\frac{i}{\hbar}\left|\int_0^T L\,dt\right|_p\right) =$$

$$\exp\left(\frac{i}{\hbar}\left(\left|Re(n)\left[\sum_{j=0}^{n-1}\frac{m}{2}\frac{\{x_{j+1}-x_j\}^2}{T}\right] - TRe\left[\frac{1}{n}\right]\sum_{j=0}^{n-1}\int_0^1 V(x_j+(x_{j+1}-x_j)s)\,ds\right|^p\right.\right.$$

$$+ \left|Im(n)\left[\sum_{j=0}^{n-1}\frac{m}{2}\frac{\{x_{j+1}-x_j\}^2}{T}\right]\right.$$

$$\left.\left.- TIm\left[\frac{1}{n}\right]\sum_{j=0}^{n-1}\int_0^1 V(x_j+(x_{j+1}-x_j)s)\,ds\right|^p\right)^{1/p}\right)$$

---

[3] Technically, as the derivative of the exponential is discontinuous in V, we should define the path contribution formula as an analytic continuation to the case that $V \neq 0$. This analytic continuation still yields the standard contribution formula for real N, and it is consistent with the exponential for

$$sgn\left(Re(n)\left[\frac{m}{2}\frac{\{x_{j+1}-x_j\}^2}{T}\right] - TRe\left[\frac{1}{n}\right]\int_0^1 V(x_j+(x_{j+1}-x_j)s)\,ds\right) =$$

$$sgn\left(Re(n)\left[\frac{m}{2}\frac{\{x_{j+1}-x_j\}^2}{T}\right]\right)$$ (and likewise for the imaginary part). The definition for the contribution formula is determined by extending the power series for the exponential to all V. The power series, the definition of the contribution formula, and the original exponential are exactly equal for arbitrarily small V.

Inserting this formula into (1) in the case of non-zero potential V, expanding the contribution formula in a Taylor series in $-\sum_{j=0}^{n-1}\int_0^1 V(x_j + (x_{j+1}-x_j)s)ds$, and simplifying, we obtain

$$\lim_{n\to\infty}\left(K - \frac{i}{\hbar}B\int\int\cdots\int \exp\left(\frac{i}{\hbar}S[0]\right)\left(T\sum_{j=0}^{n-1}\int_0^1 V(x_j + (x_{j+1}-x_j)s)ds\right)dx_1\ldots dx_{n-1} + \cdots\right)$$

In which the ellipsis denotes the sum of the other terms, S[0] is the free particle action, and B is

$$B = \left(Re(n)Re\left(\frac{1}{n}\right)|Re(n)|^{p-2}\right.$$
$$+ Im(n)Im\left(\frac{1}{n}\right)|Im(n)|^{p-2}\right)(|Re(n)|^p$$
$$+ |Im(n)|^p)^{\frac{1}{p}-1}$$

To derive an analogue of the Schrodinger equation, as in 3.B, we need only consider infinitesimal time-translations T. For the limit of small T, as in 3.B, we have

$$\lim_{n\to\infty} K - \frac{i}{\hbar}B\int\int\cdots\int \exp\left(\frac{i}{\hbar}S[0]\right)\left[\exp\left(T\sum_{j=0}^{n-1}\int_0^1 V(x_j + (x_{j+1}-x_j)s)ds\right) - 1\right]dx_1\ldots dx_{n-1} + \cdots$$

where the substitution of the exponential is accurate for small T. Furthermore, for such small time-translations, we may make the further simplification

$$\lim_{n\to\infty} K - \frac{i}{\hbar}B\int\int\cdots\int \exp\left(\frac{i}{\hbar}S[0]\right)\left[\exp\left(T\sum_{j=0}^{n-1}V(x_a)\right) - 1\right]dx_1\ldots dx_{n-1} + \cdots$$

as the value of the potential V is nearly exactly $V(x_a)$, over the entire path, for small T. Following the general procedure of Section II, and equations A.1 and A.2, now with the appearance of the sum $\sum_{j=0}^{n-1} T\, V(x_a)$, we have

$$\lim_{n\to\infty} K - \frac{i}{\hbar} B(K\exp(nTV) - K) + \cdots$$

Which, for small T, yields

$$\lim_{n\to\infty} K - \frac{i}{\hbar} Bn(KTV) + \cdots$$

Producing[4]

$$K - \frac{i}{\hbar} DTVK + \cdots \quad (10)$$

where V denotes the nearly uniform value of the potential along the infinitesimal path (equal to $V(x_a)$ above). In this case,

$$D = \exp(\text{it})\{|\cos(t)|^p - |\sin(t)|^p\}(|\cos(t)|^p + |\sin(t)|^p)^{\frac{1}{p}-1} \quad (11)$$

Given the formula (9).

B. The Generalized Schrodinger Equation

Relevant to the construction of the differential equation of the wavefunction is the behavior of the wavefunction over infinitesimal time-translations, and therefore the propagator over small T. Thus, let us consider the case of an arbitrarily small time-interval $T = \xi$. In this case, the approximations of 3.A are accurate, and the propagator is given by (10),

$$K - \frac{i}{\hbar} D\xi VK + O(\xi^2)$$

---

[4] The appearance of the n multiplying B is a result of (A. 1), which converts the sum over n terms V into the power n, producing $\exp(nTV) - 1$ which resolves into nTV for small T.

or,

$$K \exp\left(-\frac{i}{\hbar} D\xi V(x_a)\right)$$

Use of variable substitution, power series expansion, and the definition of the wavefunction in terms of the Kernel, as Feynman applies similarly in [2], suffices to demonstrate that

$$-\frac{\hbar^2}{2mz}\frac{\partial^2 \psi}{\partial x^2} + DV\psi = i\hbar \frac{\partial \psi}{\partial t} \quad (12)$$

with z given as in (9) and D as in (11).

Explicitly, the generalized Schrodinger equation is given as

$$-\frac{\hbar^2}{2m|\exp(i\mu)|_p}\frac{\partial^2 \psi}{\partial x^2} + \{|\cos(\mu)|^p$$

$$- |\sin(\mu)|^p\}(|\cos(\mu)|^p + |\sin(\mu)|^p)^{\frac{1}{p}-1}V\psi$$

$$= \exp(-i\mu)\, i\hbar \frac{\partial \psi}{\partial t}$$

with the parameter μ running from π to 2π.

This is the extended equation for the wavefunction, resulting from the generalized propagators that are derived from an analytic continuation of the path integral procedure. This is simply the standard Schrodinger Equation, but with a complex multiplier z appearing in front of the mass, and a factor D multiplying the potential. z is constrained to the particular region illustrated in **Fig. 1.** We, again, interpret these generalized versions of the Schrodinger Equation as corresponding to different universes with different associated laws of physics.

It should be noted that Feynman's process for deriving the Schrodinger Equation only applies in the case where z has a positive imaginary part.

IV. Difference with the Complex-Time Path Integral

In particle physics it is generally standard procedure to execute propagators over imaginary or complex lengths of time (e.g. [3]). The purpose of this is to ensure that certain mathematical expressions within the path integral are well-defined. We note that our unique analytic continuation procedure is entirely different from the mathematics of path integrals in imaginary or complex time. Rather than extending the time interval to the complex numbers, we are allowing the *index n itself*, a fundamental part of Feynman's formula, to be complex. Thus, rather than applying the basic laws of quantum mechanics to an imaginary length of time for the purpose of mathematical simplification, we have generalized *the laws of quantum theory themselves* and thereby produced a description of different universes where particles abide to these different laws.

Additionally, we see that our results are quite different from those of quantum theory in imaginary-time. We saw how our result for the free particle constituted a complex power of the standard propagator, with this complex exponent only appearing within a specific region in the complex plane as shown in **Fig. 1**. No such result is connected to path integration in imaginary time. The notion of the Euclidean action, which appears in path integration over an imaginary length of time, is well described in much scientific literature. We see that our equation, derived in (12), is noticeably different from that produced from the Euclidean action, in particular due to the D term, which modifies the role of the potential.

It is also worth noting that our generalized propagator (8) is equivalent to the standard equation, but with the mass M replaced by Mz.

It is important to note that we still interpret our particles as having a mass M, but the wavefunctions of the particles evolve *as if* they had a mass Mz. Analysis of particles of complex mass has been used before in physics, notably in the modeling of certain unstable particles (e.g., see [1]). A particle with decay rate E may be *formally modeled* as having a "mass" M+iE, with M the mass in the usual sense. Our Mz formula is different, as it applies to *all* particles in a network of universes corresponding to different z, rather than just *some* unstable particles in one universe. Additionally, z is a complex number *in a specific region of the complex plane*, within the region described in **Fig. 1** (parameterized by (8)); the complex quantity M+iE, however, does not have this precise sort of constraint. Furthermore, of course, the D factor multiplying V in (12) yields particle behavior distinct from simply a complex mass, in the case of the presence of a potential. In conclusion, this multiverse hypothesis yields laws of physics very different from anything studied before.

Intriguingly, we could produce the same generalized equation (12) by certain manipulations involving complex time, namely by

    1. Considering a multiverse, in which the path integral contribution formula corresponding to each universe includes a complex number w *multiplying the time interval T. All values of* w *together comprise this multiverse.*

    2. "Amending" this contribution formula, by taking the p-norm of the action, as in Section 1.B.

    3. Multiplying the mass, within the contribution formula, by w, *and the potential V by 1/*w. *Each different contribution formula yields different particle propagators, producing the*

generalized Schrodinger equation appearing throughout the multiverse as expressed in (12).

This process for producing a multiverse whose physical law is delineated by (12) does not require the use of a complex discretization index. Nonetheless, the steps (1) – (3) are entirely arbitrary and unmotivated, whereas our generalization of path skeletonization is a natural extension of quantum mechanics. The consideration of all complex scalings of the time interval, as in step (1), is unmotivated, whereas the consideration of different limits to infinity of the skeletonization index constitutes a natural extension of path integration. Furthermore, the introduction of the p-norm as in step (2), in order to protect the nature of the path integrand as a phase factor in the midst of the generalization, is a novel conception of the present paper. Crucially, however, the multiplication of the mass by w and the potential by $1/w$ is a unique property of the present generalization. The effective mass multiplication is caused by the process of integration over a complex number of variables, which can produce a complex result even if the integrand is real, as in (A.1). The $1/w$ term is a result of the fact that the sum containing the potential consists of a complex number of terms; the integration over a complex number of variables, and the computation of the resulting limit, produce a complex result, even though the original action is interpreted to be real, as in (7).

V. Separation of the Generalized Schrodinger Equation and Unitarity

In order to obtain useful physical models for nonrelativistic particle behavior, it is standard to separate the Schrodinger Equation so as to derive the time-independent Schrodinger Equation as well as the fundamental time-dependence expressions. We shall apply the same process here. Given (12),

$$-\frac{\hbar^2}{2mz}\frac{\partial^2 \psi}{\partial x^2} + DV\psi = i\hbar\frac{\partial \psi}{\partial t}$$

we set the wave function as an expression of the form X(x)T(t), with X,T functions of space and time, respectively. This will allow us to determine the basic separated solutions, and permit us to express other wave functions as combinations of these solutions. Multiplying through by z, inputting our form for the solutions, dividing by XT, and setting both sides equal to a constant, we find

$$-\frac{\hbar^2}{2m|\exp(i\mu)|_p}\frac{\partial^2 X}{\partial x^2} + \{|\cos(\mu)|^p$$

$$- |\sin(\mu)|^p\}\,(|\cos(\mu)|^p + |\sin(\mu)|^p)^{\frac{1}{p}-1}VX = EX$$

as our time-independent equation and

$$T(t) = \exp\left(-\frac{i}{\hbar}E\exp(i\mu)\,t\right) \quad (13)$$

as our time-dependence expression, with μ again constrained between π and 2π, and E a constant.

The time-dependence term (13) immediately yields issues with unitarity, as no normalization constant could ensure the conservation of probability. While this may seem a severe problem of the model, we argue that conservation of probability is *not*, in general, required for physical interpretation. For example, suppose 100 particles are released independently in a scientific apparatus. In "normal" quantum mechanics, we would calculate a 100 percent probability that we would find every particle *somewhere* in space, as "normal" quantum models are generally unitary. However, our time-dependence factor, which decays over time, clearly shows that our states cannot be normalized, generally speaking.

This is not meaningless; in our scenario, it means that some of the 100 particles *disappear* over time. If probabilities add up to, say, thirty percent, this means that we will find on average about thirty particles at the end of the experiment. Thus, because our probabilities generally decay over time, this means that particles have the capacity to vanish completely. This is an unusual property, but is *not* physically meaningless, and we could imagine the present multiverse model as logically consistent, despite the lack of unitarity in most universes.

Conclusions and Possible Future Research

We generalize the process of path integration in quantum mechanics to the case of a complex-valued time-discretization index. Subject to an augmented definition of the action, to maintain the status of the contribution formula as a pure phase factor, this generalization leads to multiple different possible propagators, due to the fact that the limit at infinity of the index is not unique. These different propagators are postulated to apply to different universes. In this framework, the solution for the free particle, as well as a particle in a potential, is analyzed, while a generalized Schrodinger Equation is developed. The differences between this approach, and path integration in imaginary time, is emphasized, as this generalization involves a complex-valued time-slicing index as opposed to time-interval. Additionally, we offer a physical interpretation of this multiverse model, consistent with its non-unitary nature.

Recent research in theoretical physics and cosmology has been founded on mathematical generalizations of physical law, notably including the anthropic analysis of the cosmological constant, which sees an extension of the Einstein field equations within the context of a multiverse model [8]. Additional such research has also focused on the implications of superstring theory, especially with respect to an anthropic landscape [4]. This model, which offers a natural analytic continuation of the Feynman path integral to complex time-slicing index, as well as an analogous "landscape" interpretation, could also pose applications to other fields of physics as well.

In the future, we would like to see an extension of the present model to relativistic physics, especially utilizing the path integral formulation of quantum field theory. Although such an approach would see difficulties in mathematical formulation, due to the measure problem associated with functional integrals in relativistic quantum theory, this extension could nonetheless provide novel information on the relativistic behavior of the multiverse model. Additionally, an extension of the model to quantum field theory could also provide enlightenment on its non-unitary nature [10]. Functional integrals associated with QFT, as demonstrated by Edward Witten and others [9], could find support as topological invariants in the fields of knot theory and low-dimensional topology; an analytic continuation of functional integration to complex discretization indices, as per the techniques of the present paper, and an associated interpretation within an anthropic multiverse context, could possibly prove useful to these domains of mathematics and lend additional support to this physical model. Furthermore, a cosmological anthropic analysis of universe evolution within this multiverse model could provide, along with additional theoretical support, possible experimental application [7].

Appendix: Mathematical Considerations

In the course of ascertaining the values of certain integrals over a complex number of variables, and determining the analytic continuation of path integration to a complex form of path skeletonization, we have used a number of assumptions about the behavior of certain expressions that are not explicitly defined. The purpose of the present appendix is *not to provide a mathematically rigorous formulation* of this continuation (which appears impossible, as none of these expressions can be explicitly delineated, when subject to a complex number of terms or variables), but to provide a consistent framework for calculating quantities relevant to path integration in the case of complex N. We might view this enterprise as similar in spirit, for example, to the notion of calculating derivatives of fractional order, but with necessarily less mathematical rigor. Ultimately, to effect an analytic continuation of this intricacy, we must decide which properties to preserve from the case of real N, and which to discard; we shall see that preserving the additive nature of integrals requires forfeiting some aspects of the real case. This particular framework, built up from a few postulates concerning the behavior of these expressions, is derived from natural assumptions about the nature of certain expressions, generalized from the real case.

The Dirac delta function, although not explicitly defined at every point in its domain, does possess a well-defined integral. Similarly, although the action (and its exponential) is a "function" of a complex number of variables in the case of complex N, and thus not truly defined, its integral is. Our key postulate in this regard is the *separation of variables:*

$$\int\int\cdots\int\prod_{j=1}^{n} f(x_j)\, dx_1 \ldots dx_n = \left(\int f(x)\, dx\right)^n \quad (A.1)$$

This separation allows us to determine an explicit value from the integral of a "function" that is not truly explicitly defined[5]. As such, as is clear in the computation of the free particle propagator, it allows us to determine a real result for a physical quantity.

In addition, as seen in Section II, we also postulate that variable substitution is possible in such multiple integrals; in particular, due to the importance of this type of integral in the physical theory, we postulate that

$$\int \ldots \int f\left(\sum_{j=1}^{n} g(x_j - x_{j-1})\right) dx_1 \ldots dx_{n-1} =$$
$$\int \ldots \int f\left(\sum_{j=1}^{n-1} g(z_j) + g\left(x_n - x_0 - \sum_{j=1}^{n-1} z_j\right)\right) dz_1 \ldots dz_{n-1} \quad (A.2)$$

provided that the limits of integration are infinite, and that f and g are explicitly defined analytic functions; this type of variable substitution is clearly allowed in the real case, and the specific integrals involved are particularly important to the physical theory.

These postulates, generalized to the complex n, constitute natural extensions of properties clearly true for integer n. In addition to postulating that the real or imaginary part of a sum is the respective sum of the real or imaginary parts of the terms, we also define the p-magnitude as

$$|A|_p = \left(\sqrt{(Re(A))^2}^p + \sqrt{(Im(A))^2}^p\right)^{1/p} \quad (A.3)$$

---

[5] This formula, in addition to allowing for the free-particle propagator to be determined, also serves to reduce the potential perturbation expansion to a calculable form, producing the exp(it) factor appearing in D (deriving from the transformation from a *sum* over a complex number n of terms, each being the value of the potential V, to a *power* of n, as a result of the integral formula).

Here, we assume that any sum of positive real terms is positive[6], and that $\sqrt{\mu^2}$ is equal to μ for μ positive. In addition, to effect the construction of the physical theory, we must determine a distributive property of complex-order sums, namely that

$$\sum_{j=1}^{n} a \cdot g_i + b \cdot h_i = a \cdot \sum_{j=1}^{n} g_i + b \cdot \sum_{j=1}^{n} h_i \quad (A.4)$$

for a, b explicitly defined constants[7].

The integration postulate (A.1) (along with the postulate concerning variable substitution, expressed in (A.2)), the summation postulate (A.4), the definition of the general p-norm in (A.3), notions of positivity, the linear nature of real and imaginary parts, and the notion that analytic functions $f(z)$ can be expanded in a Fourier integral representation (and that the exponential of a sum is the product of the exponentials) as in Section II, are the only postulates needed to define the natural analytic continuation of path integration to complex discretization index. As such, although a variety of unusual properties concerning real parts and integrals were overviewed in this Appendix, all arise from these simple postulates. (A.1) particularly, along with a few reasonable assumptions about limits[8], serves to define the value of the

---

[6] If we assume that we can "separate" any sum as in $\sum_{j=1}^{z-1} a_j = \sum_{j=1}^{n} a_j + \sum_{j=1}^{z-n-1} a_{j+n}$, for n integer, we arrive at a contradiction, for example, in the case that a_j is a positive constant, $z - 1$ is a positive integer, and $z - n - 1$ is negative, as this relation assigns a negative value to $\sum_{j=1}^{z-n-1} a_{j+n}$, a sum of positive terms. Thus, to maintain consistency, we cannot assume that the above relation works when z is complex; that is, the first or last integer number of terms cannot be separated from the sum. Indeed, there is no reason to expect that it is even meaningful to refer to the "first" term of a complex number of summands

[7] We also postulate that $Re(\sum_{i=1}^{n} a_j) = \sum_{i=1}^{n} Re(a_j)$, and similarly for the imaginary part, as described above

[8] Namely, that limits are linear, permitting the decomposition of Section III into terms of different order in T. Also, we must make the obvious assumption that limits of summation can be re-indexed as in $\sum_{i=0}^{z} a_i = \sum_{i=n}^{z+n} a_{i-n}$ for integer n.

Feynman path integral in the case of a free particle, as well as the form of the generalized Schrodinger equation, as in Sections II and III.

Analytic continuations like the present case are not unknown in theoretical physics, particularly in the domain of quantum field theory. As is well known, the calculation of the Casimir Force between two metal plates rests upon a perturbation expansion, with the terms diverging as in $\sum_{i=0}^{\infty} i$; by generalizing the behavior of the Riemann zeta function known to hold for converging series, to the case of the divergent expansion, it was concluded that the result was actually *negative*, leading to an expression for the force confirmed by experiment [6]. Although such reasoning was certainly by no means rigorous, it was eventually placed on a mathematically accurate foundation. Similarly, the present analytic continuation generalizes the behavior of integrals, sums, and real parts, as relevant to physical systems, to the case of complex n, producing results (like an integral of a real-valued function being complex) that are as unusual as a divergent series of positive terms producing a negative "result." In both cases, an analytic continuation serves to generalize known relations to a new realm, where unusual phenomena occur; the simplest aspect of the analytic continuation is present in (A.1).

These results seem strange, but sums of a complex number of terms are expected to behave in ways radically different from "explicitly defined" numbers. Indeed, it is impossible to ascribe a particular value to any of these expressions, and we therefore must use generalizations of the behavior of "usual" numbers to determine the mechanics of applying real or imaginary parts, or integrals over a complex number of variables. In a vague sense, we might view these expressions as "fuzzy," without an explicitly defined value, and the act of integration, as in (A.1), serves to establish an actual value. This is not unlike the case of the Dirac delta function, which, although undefined over its domain, still yields an explicitly defined integral.

References

[1] A Denner, S Dittmaier. "The complex-mass scheme for perturbative calculations with unstable particles." Nuclear Physics B (2006).

[2] Feynman, Richard. Quantum Mechanics and Path Integrals. New York: McGraw-Hill, 1965.

[3] Hideaki Aoyama, Toshiyuki Harano. "Complex-time path-integral formalism for Quantum Tunneling." Nuclear Physics B (1994).

[4] Raphael Bousso, Roni Harnik. "Entropic Landscape." Physical Review D (2010).

[5] Ralph Howard, Anton Schep. "Norms of Positive Operators on Lp-Spaces." Proceedings of the American Mathematical Society (1990).

[6] Sitenko, Yu. "Influence of Quantized Massive Matter Fields on the Casimir Effect." Modern Physics Letters (2015).

[7] Weinberg, Steven. "Anthropic Bound on the Cosmological Constant." Physical Review Letters (1987).

[8] Weinberg, Steven; Martel, Hugo; Shapiro, Paul. "Likely Values of the Cosmological Constant." The Astrophysical Journal (1998).

[9] Witten, Edward. "Topological Quantum Field Theory." Communications in Mathematical Physics (1988).

[10] Witten, Edward. "Quantum Field Theory and the Jones Polynomial." Communications in Mathematical Physics (1989).

[11] Yu-ichi, Takamizu; Kei-ichi Maeda. "Bubble Universes With Different Gravitational Constants" (2015).

[12] Zhengyi Lu, Yong Luo, Yizheng Hu. "Analytical Solution of the Linear Fractional Differential Equation by Adomian Decomposition Method." Journal of Computational and Applied Mathematics (2015).

A One-Electron Theory of Nonrelativistic Quantum Temporal Dynamics

Abstract:

We generalize the framework of nonrelativistic quantum mechanics, which treats time as a parameter rather than an operator, by promoting standard time to an operator and, as a consequence, introducing a two-dimensional temporal system. As a result, quantum states over time as well as space are considered, and a quantum theory of temporal dynamics is developed. This framework is used to support a new version of the Feynman one-electron universe hypothesis, by postulating the existence of large potential barriers in the past and future that effect the repeated temporal oscillations of a single particle, thus producing all particles in the universe as a consequence. A simple nonrelativistic, one-electron model is used for this quantum temporal dynamics.

I. Introduction

The Nature of Time: Philosophy and Physics

The nature of time is the greatest unsolved problem in philosophy and theoretical physics. Indeed, temporal progression is even more fundamental to the structure of the universe than spatial extent. The passage of time determines the evolution of all physical systems, and its regularity and continuity are the source of order in the cosmos. Yet, despite the centrality of time in the framework of reality, its ultimate nature remains a mystery. Is time an emergent property of a hidden structure, or is it irreducible? Is the progression of time generally linear, or is it subject to fluctuations and other quantum phenomena? Is travel to the past a possibility? In the quest to answer these deep questions, philosophy is increasingly relying upon theoretical physics and cosmology, whose insights are revealing the structure of the universe. The present paper explores a novel physical theory of time, and the role of quantum mechanics in its passage.

Classical mechanics treats time as an absolute parameter, not as an emergent property of some deeper structure. As such, while it explains the motion of mechanical bodies through space, it assumes a linear, unchanging form of *temporal* motion. This model of time is central to the "clockwork universe" of Newton, which sees the future history of the universe as determined from a set of differential equations and initial conditions[6]. Once the position and momentum of each element of a closed system is known, the future positions of each such element are determined by the mathematical formalism of classical mechanics. Quantum mechanics discards such spatial determinism, modelling particles through wavefunctions over physical space. The formalism of quantum theory replaces spatial determinism with spatial uncertainty, but maintains the absolute progression of time, in its nonrelativistic incarnation. If space is an uncertain feature of the quantum world, should the temporal continuum also be subject to quantum uncertainty?

The framework of special relativity, and the clockwork formalism of classical mechanics, lends credence to the *eternalist* theory of time, that the entirety of the future and past history of the universe exists as a single entity[1]. As such, temporal progression is an illusion of experience, an intrinsic property of a certain observational viewpoint within the spacetime structure, rather than an external characteristic. The probabilistic nature of quantum mechanics, and particularly the uncertainty of quantum state collapse, suggests the *growing block* theory, that only the past and present "exist" as such. It is indeed this phenomenon of state collapse that divides eternalism from quantum mechanics. Should the many-worlds interpretation prove accurate, that wavefunction collapse is an illusion induced by the observer's limitation, eternalism would prove compatible with quantum theory.

Temporal Nonlinearity and Temporal Emergence

The structure of nonrelativistic quantum mechanics implies certain violations to the traditional notion of temporal causality. The "delayed choice quantum eraser" experiment, a variation of the classic "double-slit" apparatus conceived by Marlan Scully, suggests that information from future events can affect the present[20]. Nonetheless, analysis of the thought experiment indicates that such violations of causality are consequences of the random behavior of quantum state collapse to an eigenvector, and that the many-worlds interpretation would prevent such retrocausality phenomena. While the linearity of the state evolution equations preserves the determinism of classical mechanics, the probabilistic nature of eigenvector collapse indeed drives many of the seemingly paradoxical characteristics of quantum systems.

General relativity, alternatively, allows causality violation through closed timelike curves, which allow superluminal communication and therefore retrocausality[12]. While locally the passage of information is never superluminal[18], the curvature of the spacetime continuum allows global phenomena to violate causal relationships and the standard sequence of time[15]. Richard Gott's solution of spacetime near a cosmic string allows a hypothetical traveler to interact with their past self, after travelling near the string[13]. Generally, however, such solutions to the Einstein Field Equations permitting time travel require negative energy densities, which have not been observed in nature[2].

The question of whether closed timelike curves exist depends on the nature of quantum gravity, which continues to elude the theoretical physics community. Different combinations of quantum mechanics and general relativity, including various models of quantum field theory in curved spacetime, suggest different answers to this question[10]. Some theories of quantum gravity, particularly loop quantum gravity, propose that the temporal continuum is an emergent property of some more fundamental structure. LQG proposes that time, although linear on macroscopic scales, is a result of a Planck-scale discontinuous topological structure. Other such models, including twistor theory, also suggest the emergence of time from a Planck-scale system[4]. Indeed, the property of background-independence is a coveted aspect of quantum gravity theories[5].

Novel Theories of Temporal Structure

The cyclical nature of time is a repeated theme in various models of cosmology. Roger Penrose's Conformal Cyclical Cosmology proposes that absolute temporal durations are irrelevant in both the entropic future state of the universe, and the high-energy initial singularity, and as such, that the geometry of the universe at these times is primarily a conformal one[16]. By using this conformality of time, he scales spacetime so as to continuously identify the future entropic history with a big-bang-like beginning, producing repeated cycles of time[17]. A variety of cyclical cosmological models, including the "big splat" model of M-Theory and the "big bounce" theories of Teller, also suggest that temporal progression has a cyclical nature[8].

John Wheeler, in a communication with Richard Feynman, developed a remarkable cyclical model of time, proposing that every electron in the universe is actually the *same* electron, moving backward and forward through time[7]. To an observer at any specific time, whose spatial slice intersects spacetime, it would appear as if multiple electrons are present. Such a model of temporal progression, in which the passage of time itself can reverse directions, suggests a notion of "temporal mechanics," that movement through time can be understood in the same way as motion through space. Such a framework would undoubtedly require additional temporal dimensions, a notion that has been explored in certain models of quantum gravity. M-Theory, the hypothetical framework unifying the different string theories, might involve hidden dimensions of time, compacified on a small scale[3]. If such additional temporal dimensions exist, the passage of time is far more complex a phenomenon than standard theories would suggest.

The mathematical formalism of nonrelativistic quantum mechanics, despite introducing uncertainty in spatial position, nonetheless upholds the absolute motion through time of classical mechanics. While a particle can exist in a superposition of distinct physical locations, its temporal position is absolute, and the probability amplitude extends only over physical space. A full understanding of the progression of time, however, should rest on a physical theory that explains movement through the temporal continuum as a consequence of a dynamical equation, rather than assuming such ad hoc. By allowing quantum dynamics to include temporal progression, and considering wavefunctions over time as well as space, quantum mechanics would see its fullest incarnation. The most natural development of physical law—from spatial certainty to spatial uncertainity—would resolve temporal certainty likewise as a macroscopic consequence of a quantum system with temporal uncertainity. The present paper concerns itself with constructing a quantum theory of temporal dynamics.

II. A Thought Experiment and Its Implications

Let us consider the following hypothetical situation. An apparatus, a "time machine", is capable of temporal travel. When the experimenter activates the apparatus, it "jumps" to a future time, say three years subsequent to the experiment, remains for one hour, and "jumps" back to the time when the experimenter initiated the apparatus. To an observer internal to the apparatus, it appears as if each temporal "jump" takes one hour to complete. The experimenter places a clock in the machine, reading 3:00 AM 9/23/2070, and activates the device.

From the perspective internal to the machine, that of the clock, three hours elapse over the course of the temporal journey. Each jump, both forward and backward, takes one hour to complete, and the machine spends one hour in "normal time" in 2073. From the perspective of an external observer, on 3:00 AM on 9/23/2070, the apparatus vanishes (through some unspecified process it leaves the standard time continuum) and an identical apparatus simultaneously appears in a different location. The experimenter determines that this other machine contains the clock he prepared minutes before. Removing the clock from this machine, he discovers that it reads 6:00 AM 9/23/2070—exactly three hours ahead of its original value.

From the perspective of the experimenter, no time has elapsed between the clock entering "jumpspace" and the clock returning to 3:00 AM, 9/23/2070. Yet, the deviation between this time, and the reading of the clock, is not puzzling to him. He realizes that each jump through time, and the brief period outside "jumpspace" in 2073, requires one hour of duration from the perspective of the clock. As such, a difference exists between the "internal time" of the apparatus and the "external time" of the surrounding system. Relative to the external system, the clock appears at the moment it left. Relative to its internal time, the clock's journey requires three hours.

This thought experiment exemplifies the distinction between the "internal time" or "instrinsic time" of a particle or sub-system, and the "extrinsic time" of the universe or surrounding system. The experimenter, following his test on the clock, enters the apparatus, while an external observer watches. The observer notes, as before, that the experimenter disappears (enclosed within the apparatus) and immediately reappears. Call the initial state of the experimenter (before the jump) State A, and the subsequent state of the experimenter (after the completion of both jumps) State B. From the perspective of the observer, no time has elapsed between State A and State B. Relative to external time, or "realtime", State A and State B exist simultaneously. Yet, during both jumps and the brief interlude in "normalspace", the state of the experimenter has undoubtedly changed. As such, State A and State B, despite existing coincident in realtime, are distinct. This experiment reveals that states evolve relative to "intrinsic time", and not relative to "external time".

The Aboriginal Australians conceived of another dimension of temporal duration[11], a "dreamtime" or "time outside of time". In a fanciful spirit, we shall adopt this nomenclature to describe "intrinsic time". The states of entities evolve relative to dreamtime, but their relation with an external system is defined by realtime. If the backwards jump of the apparatus is greater than the forwards jump, so that State B exits the apparatus before State A enters, the experimenter can interact with his past self, as both exist at the same position in realtime. As such, causality exists relative to realtime, but not relative to dreamtime, for the future and past dreamtime states of an entity can hypothetically interact.

This distinction between "internal" and "external" time provides a framework for understanding the one-electron universe of Wheeler. The ultimate goal of our quantum theory of temporal dynamics is to describe the kinematics of temporal motion, and particularly how the realtime position of a system evolves. Such a formalism would consider the internal time of a particle or system, and determine not only the probability that the particle will be found in some spatial location as a function of time, but also the chance that the particle exists at some realtime position, as a function of dreamtime. We might imagine, for example, that the apparatus is capable of random "jumps" relative to external time, and that we determine the probability that the clock internal to the box is located at any point in external time as a function of its internal time.

Thus, the wavefunction is a function of both spatial position and realtime, changing with dreamtime. Such a "temporal wavefunction" might reflect between two potential barriers, resulting in a motion of the corresponding particle forwards and backwards through realtime. Indeed, in this model, only a single electron need exist, and this dynamical motion through time produces the apparent deluge of electrons at any specific realtime position.

III. Towards a Quantum Theory of Temporal Dynamics

The present paper endeavours to construct a quantum theory of temporal dynamics, describing the chronological "position" of a particle much in the same way that standard quantum theory treats its physical position. Much as typical quantum particles exist in a superposition of distinct physical locations, a continuous "wavefunction" over space, whose value is proportional to the associated probability amplitude, a particle in "temporal quantum theory" would exist in a continuous superposition of distinct *times*, with a nonzero probability of existing in one *time* or another. This project represents the fullest completion of quantum theory, which, while discarding the Newtonian conception of *definite space*, nonetheless upholds that of *definite time*.

It should be noted that the quantum temporal dynamics explored in this article will be nonrelativistic; that is, applying to systems with low energy, where the standard Schrodinger equation serves as a useful approximation. Of course, any complete theory of time would include the effect of relativistic phenomena, but the development of a quantum field theory of temporal dynamics extends far beyond the scope of the present paper. Such an enterprise represents the most natural extension of these results, but its implementation would involve an investigation of considerable complexity. As such, all the succeeding analysis of the theory will start by assuming nonrelativistic quantum theory, and particularly the standard Schrodinger equation and low-energy Hamiltonian. Nonetheless, some aspects of special relativity, and particularly the form of the Minkowski metric, will be considered.

The centerpiece of QTD is an equation describing the evolution of "temporal wavefunctions", which take as argument spacetime position and yield the probability amplitude (or weighting coefficient in the superposition) that the particle will be found in that state. Clearly, as the thought experiment of the preceeding section demonstrated, an additional variable must be included, with respect to which these temporal wavefunctions evolve. This variable, which we called "dreamtime", could be considered a sort of "intrinsic time" relative to the particle, rather than the "extrinsic time" of the larger system (or universe). By providing a mathematical mechanism for describing how temporal states evolve, we not only make the remarkable proposition that particles can exist in a superposition of different times, but also that "movement through time" is a mechanical quality. That is, we propose that the movement of systems through time can be analyzed in much the same way as their movement through physical space, and that this movement is subject to quantum phenomena (e.g., fluctuations and "jumps"). Much as quantum theory reduces to the standard classical formalism for macroscopic systems, our QTD would reduce to the standard, linear progression through time for macroscopic phenomena. However, on the Planck scale, nontrivial "temporal mechanics" might exist.

As we shall consider in Section VII, QTD lends credence to the "one-electron" hypothesis of John Wheeler, by suggesting that every electron in the universe (or quark, muon, etc.) is actually the *same* electron, but moving back and forth through time, so to the observer at any *specific* time, it appears as if a multitude of electrons are present. The original one-electron universe was proposed in the context of quantum field theory, but QTD instead employs nonrelativistic quantum mechanics, providing a mechanism for this electron to move through time. Roughly, QTD suggests that vast potential barriers exist at the past and future singularities of the universe, and that *the* electron bounces back and forth between these barriers, populating the cosmos. Eventually, the electron might tunnel through one of the barriers, and populate another universe; in Section VIII, this remarkable cyclical cosmology will be investigated.

To develop the central equation of QTD, we begin by analyzing the standard one-dimensional Schrodinger equation for a single particle:

$$-\frac{\hbar^2}{2m}\nabla^2\psi + V\psi = i\hbar\frac{\partial\psi}{\partial t} \quad (3.1)$$

In QTD, we replace the standard wavefunction $\psi(r,t)$ with a more general temporal form $\psi(r,t;T)$, where t is the external "realtime" and T is the "dreamtime" coordinate with respect to which ψ evolves. As such, ψ is primarily a function of (r,t), expressing the probability amplitude that the particle will be found at (r,t), and the form of this function changes with dreamtime T. The most natural way to generalize (3.1) to the case of a temporal wavefunction $\psi(r,t;T)$ is to treat (r,t) itself as a spatial coordinate, as it serves an analogous purpose as $r = \{x,y,z\}$ in standard quantum mechanics, and treat T, the parameter effecting wavefunction evolution, as the temporal variable. Now, the del-squared (Laplacian) operator on the Minkowski space of four-dimensional spacetime (r,t) is the d'Alembertian, given as

$$\Box = \nabla^2 - \frac{1}{c^2}\frac{\partial^2}{\partial t^2} \quad (3.2)$$

With c the speed of light in vacuum; generalizing (3.1), replacing the spatial coordinates with the four-dimensional (r,t) and time by T, we have

$$-\frac{\hbar^2}{2m}\Box\psi + V\psi = i\hbar\frac{\partial\psi}{\partial T} \quad (3.3)$$

This represents a "first attempt" at creating a consistent mathematical equation for the evolution of Dreamtime states. Unfortunately, as Section VI describes, equation (3.3) is inconsistent with the standard Schrodinger equation, and would produce anomalies (like temporal accelerations) that have not been observed in nature. In order to achieve consistency, (3.3) must be modified by adjusting the formula for the d'Alembertian operator. Specifically, we must replace the speed of light c with a free parameter a, producing the "modified d'Alembertian"

$$\Box_a = \nabla^2 - \frac{1}{a^2}\frac{\partial^2}{\partial t^2} \quad (3.4)$$

The value of a, the "temporal velocity", is unspecified by the theory itself, but Section VI estimates its value given experimental data. Thus, our modified quantum equation is

$$-\frac{\hbar^2}{2m}\Box_a \psi(r,t;T) + V\psi(r,t;T) = i\hbar \frac{\partial \psi(r,t;T)}{\partial T} \quad (3.5)$$

describing the evolution of temporal quantum states in dreamtime T. The natural next step is to consider the form of the potential V.

IV. The Potential in Quantum Temporal Dynamics

The Underlying Potential Formula

Typical notions of causality do not apply in QTD. Previous attempts to construct theories involving multiple time dimensions determined that interactions between entities are only possible if both entites exist at the same two-time coordinate $(t;T)$. Quantum Temporal Dynamics, although a theory involving two temporal dimensions (dreamtime and realtime) is completely different, as it treats these dimensions in a distinct fashion. Realtime is incorporated into the d'Alembertian operator with the spatial coordinates, while dreamtime has the same role as standard time in the traditional Schrodinger equation. Realtime is an external coordinate indexing the position of a particle relative to the larger system, while dreamtime evolves the state of that particle. As our thought experiment demonstrated, causality exists relative to realtime, not dreamtime.

If a particle's temporal wavefunction evolves in T so that it returns to a position t that it occupied at an earlier value of T, to the external observer at t, it appears as if two particles are present. These "particles" can clearly interact, e.g. through electrostatic forces. As the previous section demonstrated, self-interaction is possible, if a particle's future and past dreamtime selves coincide at the same position in realtime. As such, realtime indexes causal positions, while temporal evolution itself is effected through dreamtime. If dreamtime and realtime do not coincide, apparent violations of causality can occur.

Such a model explains the one-electron hypothesis, as the evolution of the single electron in dreamtime (reflecting between two potential walls) produces an apparent deluge of electrons at any realtime position t, being the incarnations of the particle at distinct dreamtime positions. Despite existing at different dreamtime points, these particles clearly can interact. In the remainder of the present article, we shall consider a nonrelativistic universe consisting only of electrons, and analyze how QTD can explain the characteristics of this model universe. Further development of QTD should consider interactions between distinct particle types (e.g. quarks, leptons, etc.) in a relativistic framework.

We open our analysis of this one-electron universe by considering the form of the potential V. The potential consists of two central components, the enormous potential walls at the beginning and end of the universe, and the electrostatic interactions (Coulomb force) between the consistuent particles of the universe (we simplify the model by neglecting other contributions to the potential). The former is clearly a function of the form $R(t)$ of realtime, being zero for most of realtime and adopting a high value for two values of realtime. The nature of this function will be explained in Section VII. The particle interactions constitute the self-interaction of the electron with itself, as its states at different dreamtimes conincide in realtime. Denoting the self-interaction by $S(r,t)$, we have that the potential is given as

$$R(t) + S(r,t) \quad (4.1)$$

Now we consider the form of the self-interaction $S(r,t)$. Clearly, this potential is a sum over contributions from states of the electron at distinct dreamtimes, that coincide in realtime. The obvious form of $S(r,t)$ is an integral over T', of the potential at (r,t) generated by the electron at T' (whose state is $\psi(r',t';T')$). That integrand itself is an integral, over (r',t'), of the contribution to the potential at (r,t) by the portion of the wavefunction at (r',t'). The most natural way to derive this contribution is to consider the probability amplitude itself as a charge density, and thus multiply by $|\psi(r',t';T')|^2$.

This contribution is proportional to $V(r-r')$, the potential between charges at r and r'. Clearly, states significantly separated in realtime do not interact, but as the temporal wavefunctions will have some uncertainty ("fuzziness") in realtime position, we should allow for some slight violations in realtime causality. We account for such by multiplying the above contribution by $\delta_\varepsilon(t-t')$, a Gaussian with width ε that highly weights contributions from states close in realtime.

Considering the above argument, we modify (4.1) to produce the formula

$$R(t) + \iiint |\psi(r',t';T')|^2 V(r-r')\delta_\varepsilon(t-t')dr'dt'dT' \quad (4.2)$$

for the potential at (r,t). We imagine that the width of typical temporal waveforms in realtime is on the Planck scale, a level unobservable by present technology, and as such, that the "fuzziness" of these quantum states would presently go unnoticed. We make this proposal, as the Planck scale seems to be the level at which novel physical phenomena emerge. Thus, the scale of ε, the width of the Gaussian that sets the level of realtime causality violations, is roughly the Planck time.

To prevent particles from interacting with their present dreamtime state, we stipulate that the outermost integral over dreamtime should exclude some small interval $(T-\lambda, T+\lambda)$, with λ on the scale of the Planck time. $V(r-r')$ itself is a form of potential "density", with different units than the standard Coulomb potential. $R(t)$, explained in Section VII, constitutes the enormous potential walls at the past and future singularities of the universe. (4.2) is the foundation of QTD's analysis of the potential.

Deriving the Known Formula

Now that the general form of the potential in QTD, (4.2), is known, it seems natural to make sense of it, and how known physical phenomena emerge from it. We begin by analyzing the formula for the self-interaction:

$$\iiint |\psi(r',t';T')|^2 V(r-r')\delta_\varepsilon(t-t')dr'dt'dT' \quad (4.3)$$

We write the wavefunction as a separation of spatial and temporal states, $\psi(r',t';T') = \theta(r',T')\phi(t',T')$, with θ adopting the role of the standard quantum wavefunction (the probability amplitude commonly measured) and $\phi(t', T')$ a Gaussian with Planck-scale width that generally moves forward in realtime as dreamtime progresses (a wavepacket; see the succeeding section for a more thorough description). ϕ reflects against the enormous potential barriers $R(t)$, thus producing the back-and-forth motion of the electron through dreamtime. This temporal wavepacket might be moving forward or backward relative to realtime, depending on what leg of the cosmic journey the electron is currently on. We set $t'_{T'}$ to be the center of this Gaussian, so that for $t' \neq t'_{T'}$, $\phi(t', T')$ rapidly decays on Planck-level scales of $t'_{T'} - t$. Introducing this separation into the above formula, we have

$$\iiint |\theta(\mathbf{r}', T')|^2 |\phi(t', T')|^2 V(\mathbf{r} - \mathbf{r}') \delta_\varepsilon (t - t') d\mathbf{r}' dt' dT'$$

Now, δ_ε decays quickly for $t' \neq t$. Thus, only those values of T' such that $t'_{T'} \approx t$ contribute significantly to the integral. Let us define T_j' so that $t'_{T_j'} = t$. Thus, the integral over dreamtime reduces, approximately, to a sum of smaller integrals about each separate T_j' (excluding, of course, T itself). Thus, we have

$$\sum_j \iiint |\theta(\mathbf{r}', T')|^2 |\phi(t', T')|^2 V(\mathbf{r} - \mathbf{r}') \delta_\varepsilon (t - t') d\mathbf{r}' dt' dT'$$

which, employing the separation of variables, reduces to

$$\sum_j \int \left(\int |\theta(\mathbf{r}', T')|^2 V(\mathbf{r} - \mathbf{r}') d\mathbf{r}' \right) \left(\int |\phi(t', T')|^2 \delta_\varepsilon(t - t') dt' \right) dT'$$

Now, $\theta(\mathbf{r}', T')$ likely does not change significantly for small (Planck-order) changes in T'. Indeed, θ represents that standard (spatial) quantum wavefunction that typically measured, and such states usually change over the scale of atomic time, which is orders of magnitude greater than Planck time. As each dreamtime integral in the sum exists over a (roughly) Planck-scale interval, we can treat θ as being effectively constant with respect to T'. We thus have

$$\sum_j \left(\int |\theta(r', T_j')|^2 V(r-r') dr'\right) \int\int |\phi(t', T')|^2 \delta_\varepsilon(t-t')\, dt'\, dT'$$

Assuming that the width of the temporal wavepacket does not change appreciably for significant spans of dreamtime, as Section VI will demonstrate, the temporal integral above (the second factor in the summand) is largely constant, not depending on the value of realtime t. Assuming that the temporal wavepacket's associated probability density is of the form

$$|\phi(t', T')|^2 = exp(-a(t' - t'_{T'})^2) \quad (4.4)$$

a Gaussian of largely unchanging width, the integral in question can be written as

$$\int\int exp(-a(t' - t'_{T'})^2)\, \delta_\varepsilon(t-t')\, dt'\, dT'$$

with the dreamtime integral over a small interval about T_j', and the realtime one over a Planck-level interval about t. Simplifying, we have

$$\int\int exp(-a(t' - t - (t'_{T'} - t))^2)\, \delta_\varepsilon(-(t'-t))\, dt'\, dT'$$

It should be noted that, for any particular value of j, the center of the temporal Gaussian can be written as $t'_{T'} = \pm T' + b_j$, with b_j a constant depending on the leg of the electron's temporal journey, and the positive or negative sign depending on the direction of temporal progression. Employing a substitution of variables, we have

$$\int\int exp(-a(w - l)^2)\, \delta_\varepsilon(w)\, dw\, dl$$

a formula that doesn't depend on the value of realtime t. Each integral is over a small, Planck-sized interval about zero. Referring to the above integral as the constant p, the approximate formula for the self-interaction reduces to

$$\sum_j p \left(\int |\theta(r', T_j')|^2 V(r - r') dr' \right)$$

Incorporating a value of the potential density $V(r - r')$ proportional to the standard Coulomb factor, we have

$$\sum_j p \left(\int |\theta(r', T_j')|^2 \frac{Dq^2}{|r - r'|} dr' \right)$$

where q is the charge on the electron, and the constant D is proportional to the Coulomb constant k as

$$D = k/p$$

with p given as

$$p = \int \int exp(-a(w - l)^2) \, \delta_\varepsilon(w) \, dw \, dl \quad (4.5)$$

where the Gaussian is

$$\delta_\varepsilon(x) = \frac{1}{\varepsilon\sqrt{2\pi}} exp\left(-\frac{x^2}{2\varepsilon^2}\right) \quad (4.6)$$

Given this value for D, the self-interaction is approximately

$$S(r, t) = \sum_j \int |\theta(r', T_j')|^2 \frac{kq^2}{|r - r'|} dr' \quad (4.7)$$

with T_j' defined so that $t'_{T_j'} = t$. This formula yields the potential at (r, t) for a nonrelativistic, one-electron universe where electrostatic force communication is instantaneous.

Comparing With the Known Formula

Equation (4.7), which is only an approximation of (4.3), reduces to the well-known formula for the potential. To the perspective of the observer affixed to a particular point in realtime, the subject of our thought experiment, it appears as if the contribution of each different j in the sum of (4.7) derives from a separate particle. In reality, these correspond to distinct points in dreamtime of the same particle; this is simply a restatement of the notion that causality is violated in dreamtime, allowing a particle's future dreamtime self to interact with (affect) its past dreamtime self. In the one-electron universe, it is indeed this self-interaction that is responsible for all particle forces, aside from the potential walls at the ends of the universe.

Indeed, (4.7) is simply the formula for an electrostatic potential caused by a number of different particles indexed by j, each particle having the wavefunction $\theta(\mathbf{r}', T_j')$. Furthermore, as is well-known, the standard spatial wavefunction of a particle can be treated electrostatically as a charge density, with the potential caused by the particle an integral of the form

$$\int |\psi(\mathbf{r}', t)|^2 \frac{kq^2}{|\mathbf{r} - \mathbf{r}'|} d\mathbf{r}'$$

where the squared modulus of the wavefunction acts as charge density[9]. As such, (4.7) gives us a potential caused apparently by multiple different particles, each of which is the single electron on a different lap of its back-and-forth journey through time. Of course, however, this equation is only an approximation, assuming, for example, that the temporal wavepacket decays to virtually zero for values of realtime significantly different from the center $t'_{T'}$. In reality, there might be negligibly small deviations from (4.7), particularly on Planck scales.

V. Derivation of the Decoupled Equations and Temporal Duality

Given (4.2), our temporal Schrodinger equation, (3.5), becomes

$$-\frac{\hbar^2}{2m}\Box_a \psi(r,t;T) + \left\{ R(t) + \iiint |\psi(r',t';T')|^2 V(r-r') \delta_\varepsilon(t-t') dr' dt' dT' \right\} \psi(r,t;T) = i\hbar \frac{\partial \psi(r,t;T)}{\partial T} \quad (5.1)$$

Equation (5.1) is the central equation of Quantum Temporal Dynamics. All the aspects of the theory, besides the precise form of the endtimes potentials explored in Sections VII and VIII, can be derived from this equation. Let us begin by considering the solutions separated in realtime and space that were introduced in the previous section. That is, we set

$$\psi(r,t;T) = \theta(r,T)\phi(t,T) \quad (5.2)$$

describing the quantum state as a product of the spatial wavefunction (the state considered in standard quantum mechanics) and the temporal wavepacket, which typically moves forward in a linear fashion. For the time being, we simplify our analysis of the previous section, and consider a potential given as

$$V = V(r) + R(t) \quad (5.3)$$

the sum of an unchanging spatial potential and the endtimes potential. (5.3) constitutes a dramatic simplification of (4.7), where the electrostatic potential is modelled as unchanging in realtime. With this simplified potential, subsituting the separated form of the wavefunction, we have

$$-\frac{\hbar^2}{2m}\Box_a \theta(r,T)\phi(t,T) + \{R(t) + V(r)\}\theta(r,T)\phi(t,T)$$
$$= i\hbar \frac{\partial}{\partial T}\{\theta(r,T)\phi(t,T)\}$$

Simplifying, and applying the definition of $\Box_a$, results in

$$-\frac{\hbar^2}{2m}\left(\nabla^2 - \frac{1}{a^2}\frac{\partial^2}{\partial t^2}\right)\theta(r,T)\phi(t,T) + \{R(t) + V(r)\}\theta(r,T)\phi(t,T)$$
$$= i\hbar\{\theta^*(r,T)\phi(t,T) + \theta(r,T)\phi^*(t,T)\}$$

This yields

$$-\frac{\hbar^2}{2m}\left\{\phi(t,T)\nabla^2\theta(r,T) - \theta(r,T)\frac{\partial^2}{\partial t^2}\frac{1}{a^2}\phi(t,T)\right\}$$
$$+ \{R(t) + V(r)\}\theta(r,T)\phi(t,T)$$
$$= i\hbar\{\theta^*(r,T)\phi(t,T) + \theta(r,T)\phi^*(t,T)\}$$

Dividing through by $\theta(r,T)\phi(t,T)$, we have

$$-\frac{\hbar^2}{2m}\left\{\frac{\nabla^2\theta(r,T)}{\theta(r,T)} - \frac{1}{a^2}\frac{\phi''(t,T)}{\phi(t,T)}\right\} + \{R(t) + V(r)\}$$
$$= i\hbar\left\{\frac{\theta^*(r,T)}{\theta(r,T)} + \frac{\phi^*(t,T)}{\phi(t,T)}\right\}$$

Simplifying,

$$-\frac{\hbar^2}{2m}\frac{\nabla^2\theta(r,T)}{\theta(r,T)} - i\hbar\frac{\theta^*(r,T)}{\theta(r,T)} + V(r)$$
$$= -\frac{\hbar^2}{2ma^2}\frac{\phi''(t,T)}{\phi(t,T)} + i\hbar\frac{\phi^*(t,T)}{\phi(t,T)} - R(t) = D(T)$$

As functions of space and realtime are equal to each other, the only possibility is that both are equal to a constant $D(T)$ (which itself generally depends on dreamtime, but not on (r,T)). Reducing the resultant couplet of equations, we have

$$-\frac{\hbar^2}{2m}\nabla^2\theta(r,T) + (V(r) - D)\theta(r,T) = i\hbar\frac{\partial}{\partial T}\theta(r,T)$$

$$-\frac{\hbar^2}{2ma^2}\frac{\partial^2}{\partial t^2}\phi(t,T) + (-R(t) - D)\phi(t,T) = -i\hbar\frac{\partial}{\partial T}\phi(t,T)$$

Now, each resultant equation is a variation of the standard Schrodinger equation (3.1), in the latter case with a modified mass $-ma^2$. As is well known, addition of any time-varying potential that is constant over space produces equivalent solutions to the Shcrodinger equation, perhaps with an added phase. Indeed, a potential constant over space produces no forces, and therefore is physically indistinguishable from no potential at all. Thus, the two equations above are *equivalent* to

$$-\frac{\hbar^2}{2m}\nabla^2\theta(r,T) + V(r)\theta(r,T) = i\hbar\frac{\partial}{\partial T}\theta(r,T) \quad (5.4)$$

$$\frac{\hbar^2}{2ma^2}\frac{\partial^2}{\partial t^2}\phi(t,T) + R(t)\phi(t,T) = i\hbar\frac{\partial}{\partial T}\phi(t,T) \quad (5.5)$$

(5.4) is the standard Schrodinger equation, but with realtime replaced with dreamtime. As such, it stipulates that the spatial waveform evolves in dreamtime the same way that standard wavefunctions evolve in time. (5.5) describes the temporal wavepacket $\phi(t,T)$, which follows the typical Schrodinger equation, but with the potential given as $R(t)$ and the mass as $-ma^2$. We imagine, as was noted in the previous section, that this wavepacket is an extremely thin Gaussian (roughly Planck-width) that moves forward with a unitless speed of 1. This represents the typical progression through time, and when the wavepacket moves in such a way, for all intents and purposes dreamtime and realtime are equivalent, thus reducing (5.4) to the standard equation in realtime.

However, at the points in realtime where the endtimes potential $R(t)$ becomes relevant, the temporal wavepacket reflects and reverses direction.

(5.5) solidifies a notion called *temporal duality*. Namely, when atomic units are employed, the temporal evolution of a system is equivalent to the spatial evolution of a system, with the mass multiplied by $-a^2$. In this case, a speed of 1 in the temporal situation corresponds to a speed of roughly $(10^{-10}$ m$) / (10^{-17}$ sec$)$ in the spatial situation, etc. Thus, we need only consider the modified spatial system to determine the evolution of the temporal system.

Conveniently, the probability density of the wavefunction is the product of the densities of the spatial and temporal wavefunctions, as

$$\iint |\psi|^2 \, drdt = \iint |\phi(t,T)|^2 |\theta(r,T)|^2 \, drdt$$

$$= \left(\int |\phi(t,T)|^2 dt \right) \left(\int |\theta(r,T)|^2 dr \right) \quad (5.6)$$

Thus, normalization of each such function separately ensures normalization of the combined wavefunction.

VI. Derivation of the Temporal Velocity

Now we have developed the mathematical machinery to calculate the temporal velocity a, at least within an order of magnitude. We assume, as before, that the temporal wavepacket $\phi(t,T)$ has a roughly Planck-order width, that remains Planck-order for significant spans of dreamtime T. Quantum wavepackets, in accordance with the uncertainty principle, generally disperse over long periods of time. Limiting this dispersion will provide a lower bound on the value of the temporal velocity. Ultimately, with a value for a high enough, the constant speed of the wavepacket can be maintained, even in the presence of time-varying potentials that would cause significant "temporal acceleration" (changes in the speed of the temporal wavepacket) for low values of a.

We begin by employing the "temporal duality" developed in the previous section, where a temporal system is equivalent to a spatial system (in atomic units). Given the separated temporal equation in a realtime region without the endtimes potentials $R(t)$,

$$\frac{\hbar^2}{2ma^2}\frac{\partial^2}{\partial t^2}\phi(t,T) = i\hbar\frac{\partial}{\partial T}\phi(t,T) \quad (6.1)$$

in atomic units we have the moving wavepacket solution[19]

$$\phi(t,T) = \left(\frac{2c}{\pi}\right)^{1/4}\frac{exp\left(\frac{-ct^2 + i(lt + l^2T/2ma^2)}{1 - 2icT/ma^2}\right)}{\sqrt{1 - 2icT/ma^2}} \quad (6.2)$$

where c is a free parameter and

$$\frac{-l}{ma^2} = v$$

The resultant probability density is given as

$$|\phi(t',T')|^2 = \sqrt{2/\pi}\, w\, exp(-2w^2(t - lT/ma^2)^2)$$

or alternatively

$$|\phi(t',T')|^2 = \sqrt{2/\pi}\, w\, exp(-2w^2(t - vT)^2) \quad (6.3)$$

where

$$w = \left(\frac{c}{1 + (2cT/ma^2)^2}\right)^{1/2}$$

The width of the probability density Gaussian is proportional to

$$\frac{1}{w}$$

Thus, the width after a dreamtime T is given by

$$\left(\frac{1 + (2(1/d)^2 T/ma^2)^2}{(1/d)^2}\right)^{1/2} \quad (6.4)$$

where d is the initial width at dreamtime $T = 0$. Our primary concern is that the wavepacket should preserve its Planck-order width over its entire journey. As there are 10^{80} particles in the observable universe, we can infer that the electron makes at least 10^{80} laps between the two potential reflectors $R(t)$. Assuming that each such lap is on the order of 10^{10} years in duration, in atomic units, the total length of the journey in dreamtime is 10^{114} (the length in realtime is on the order of 10^{17}), before the electron possibly tunnels through one of the potential barriers (see Section VII for an analysis of this phenomenon). Thus, the dispersion of the temporal wavepacket over this journey is

$$10^{114}/mda^2$$

with d the initial width. As the Planck time is 10^{-26} in atomic units, and the electron's mass is 1, our requirement reduces to

$$10^{114}/(a^2 10^{-26}) \approx 10^{-26}$$

giving the "effective temporal mass"

$$a^2 = 10^{166} \quad (6.5)$$

in atomic units, yielding the temporal velocity

$$a \approx 10^{81} c \quad (6.6)$$

Alternatively, the uncertainty principle applying in the dual spatial system

$$(\Delta x)^2 (\Delta p)^2 \geq \frac{\hbar^2}{4}$$

provides the same limits on the size of a, for in atomic units we have

$$(\Delta x)^2 (a^2 \Delta v)^2 \geq \frac{1}{4} \quad (6.7)$$

where the temporal wavepacket dispersion is given by $\Delta v T$. As the corresponding spatial uncertainty $\Delta x = 10^{-26}$, we have that

$$\Delta vT \geq \frac{1}{2}a^{-2}10^{140}$$

applying the value $T = 10^{114}$. As this quantity should remain on the order of 10^{-26}, a should be on the order of $10^{81}\ c$.

For such a large effective mass in (5.5), any anomalous changes in the speed of the temporal wavepacket, due to time-varying potentials that appear in the equation, will be negligable. Even if the electron's temporal wavepacket is subject to a constant force of 1 Newton for its entire existence, temporal duality demonstrates that during this length of dreamtime, the speed of the wavepacket will change by a factor of

$$\frac{\left(\frac{1\ N}{10^{166}\ m_e}\frac{1}{10^{17}}s\ 10^{114}\right)}{0.01\ c} \approx 10^{-46}$$

Thus, it would take enormous time-varying potentials to cause significant changes in the speed of the electron through time. As such, temporal progression is almost always exactly linear, except at the two endtimes potentials, which are large enough (and time-varying) to accelerate the temporal wavepacket.

VII. The Endtimes Potential

Quantum Temporal Dynamics describes the mechanics of how particles move through time. As such, it permits not only the linear progression through time, but also accelerations, reflections, and other phenomena, claiming that forces and potentials can affect the passage of time; furthermore, quantum phenomena at the Planck scale can cause fluctuations or aberrations in time's passage. Nonetheless, due to the high value for the effective temporal mass (or the temporal velocity) these effects are extremely small, especially at macroscopic scales. We propose, however, that at the past and future singularities of the universe (the Big Bang and, perhaps, the Big Rip, or the conformally modified heat death of Roger Penrose's CCC), vast potentials fill the cosmos, resulting in a time-varying $R(t)$ that is zero for most of realtime and spikes to a high value at two values of realtime. These potentials are rectangular barriers, with an extremely thin base. The repeated back-and-forth motion of the electron between the barriers produces the apparent multitude of electrons at any point in realtime.

As the subsequent analysis will reveal, it appears most natural to set the width of these barriers as inversely proportional to the "effective temporal energy"

$$L = \frac{\hbar}{|E|} \quad (7.1)$$

where E is the kinetic energy of the dual spatial wavepacket. We have that, roughly, $E = \frac{-1}{2} m a^2 v^2$, where v is the unitless speed of the wavefunction through time, almost exactly 1. Thus, given the vast effective energy of the temporal wavepacket, the potential barriers are extremely thin. Formula (7.1) was chosen to allow a significant range in the size of the potential barrier to yield roughly the same probability of transmission.

We take the dual spatial situation in atomic units, with a particle of mass $m = -10^{166}$ encountering a rectangular barrier of height U. The probability of transmission (tunneling) is given as[14]

$$exp\left(-2\sqrt{m^2}L\sqrt{\frac{U}{E} - 1}\right) \quad (7.2)$$

simplifying to

$$exp\left(-2 * 10^{166} L \sqrt{\frac{U}{E} - 1}\right)$$

As the preceding section argued, the probability of transmission each time the electron encounters the future or past barrier is roughly 10^{-80}, as 10^{80} electrons are visible to an observer at any point in realtime. Thus, we have

$$exp\left(-4\sqrt{\frac{U}{E}} - 1\right) = 10^{-80}$$

yielding a potential barrier of height

$$U = -2.2 * 10^{169} \quad (7.3)$$

in atomic units, or Hartrees. Alternatively, the barrier is roughly -10^{152} Joules in height.

VIII. A Cyclical Multiverse Through Quantum Tunneling

The preceeding section offers an explanation for the one-electron universe, namely that large potential barriers reflect a particle between two values of realtime, and that roughly 10^{80} laps between these temporal positions produce all the electrons observable at any point in time. As the future and past dreamtime states of a particle can interact electrostatically if they coincide in realtime, as equation (4.7) demonstrates, these seemingly different electrons can interact, producing an apparent multi-particle universe. While 10^{90} years of dreamtime elapse before this universe is "formed" through the back-and-forth motion of the single electron, only 10^{10} years of realtime elapse between the potential barriers.

The analysis of the previous section suggests an intriguing possibility. With the height of the potential barrier we ascertained, a quantum tunneling event through either the past or future barrier is likely after 10^{80} reflections. As such, once the electron finishes populating this universe, it might tunnel through the future potential wall and populate another universe. There might be an infinite, or very large, number of potential barriers, and any interval between two such barriers would constitute a "universe". Quantum tunneling would allow the electron to penetrate from one such "universe" to another.

The calculations of Sections IV and VI assume, however, that the temporal wavepacket evolves over 10^{90} years of dreamtime; if each of the infinite number of barriers has the same height and width, then tunneling through either the past or future barrier of any given universe is equally likely, and hence the electron, after leaving our universe, would eventually tunnel back into it. Thus, the electrons in our universe would correspond to any number of distinct "classes", corresponding to disconnected ranges of dreamtime during which the electron exists in our universe. The total length of dreamtime when the electron exists, in this scenario, is very likely greater than 10^{90} years. To prevent this from happening, it seems natural to allow the height of each successive barrier to be significantly lower than the height of the previous barrier (or vice versa), to promote unidirectional movement through the realtime continuum. Furthermore, the number of universes should be large but finite, and surrounding the multiverse on the past and future realtime sides is an infinite void, through which the single electron travels indefinitely (see **Fig. 8.1**).

As each sucessive barrier is slightly lower than the previous, with a greater probability of transmission for the incident electron, universes to the future in realtime are significantly less dense than universes to the past in realtime, with a lower population of electrons. Further in the past, universes have an ever greater particle density, while further in the future, this density is ever sparser. Ultimately, beyond the past barrier of the multiverse is a single electron which tunnels through the first potential, and beyond the future potential is likewise an endless void with a single electron.

IX. Philosophical Implications

Quantum Temporal Dynamics addresses the problem of time by subjecting temporal progression to the same kinematical formalism that describes movement thorough space. Specifically, it applies the nonrelativistic framework of quantum mechanics, which involves spatial uncertainty, to time, determining the probability that a particle exists in one time or another (and evolving states of temporal superposition). To ensure the consistency of QTD, it is necessary to consider two distinct dimensions of time: the "internal time" of a single particle or sub-system – "dreamtime" – and the "external time" or "realtime" of the surrounding universe. QTD ascertains the probability that a particle exists at any given realtime, as dreamtime moves forward.

As such, the two-dimensionality of time is central to the inner workings of QTD. Temporal Dynamics implies that time, far from being the linear entity of classical mechanics, is an intricate, multi-layered beast, with distinct dimensions and components. Traditional causal relationships exist relative to realtime, as two particles must exist coincident in realtime to interact (assuming instantaneous force transmission); the quantum state of a particle, however, evolves relative to dreamtime. This framework significantly changes the meaning of temporal duration, as the two-fold nature of time (internal and system-wide) is fundamentally different from unidirectional progression.

QTD thus changes the meaning of the problem of time. Realtime suggests an eternalist perspective, namely that the entire history of realtime is pre-determined. Indeed, as the previous section suggested, the values of the potential barriers, even in the realtime future of the universe, are pre-determined (one could say that the "structure" of this multiverse is eternalist, in that it "exists" independent of any particular position in time). The perspective from dreamtime, however, suggests a "growing block" model, in that the evolution of the electron's quantum state is determined relative to dreamtime. However, the form of (4.2) upholds an eternalist framework, as the entire future history of this quantum state must be determined to consider the value of the potential at some point in realtime and some spatial location. Thus, QTD is by nature a fundamentally determinist enterprise; as the randomness of quantum state collapse disrupts such determinism, it seems necessary to unite QTD with the many-worlds intepretation.

The source of time's unexplained nature in the frameworks of theoretical physics is the assumption of unidirectional temporal progression, without consideration of temporal "kinematics". By explaining the progression of time in the same way as mechanics explains motion through space, Quantum Temporal Dynamics provides a physical understanding of time's passage. The reflections of a single particle against potential barriers explain the appearance of matter and energy in the universe. Although a speculative idea, QTD has the potential to change our understanding of the universe.

References

1. *The 3D/4D Controversy. Stanford Encyclopedia of Philosophy*, plato.stanford.edu/entries/time/#3D4Con.

2. Alcubierre, Miguel. *The Warp-Drive: Hyper-Fast Travel Within General Relativity. www.arxiv.org*, arxiv.org/abs/gr-qc/0009013. Accessed 21 Oct. 2016.

3. Bars, Itzhak. *Survey of Two-Time Physics. Arxiv*, arxiv.org/abs/hep-th/0008164. Accessed 14 Oct. 2016.

4. Biswas, S., et al. "Time in Quantum Gravity." *ArXiv. ArXiv.org*, arxiv.org/pdf/gr-qc/9906010v1.pdf. Accessed 13 Oct. 2016.

5. Carroll, Sean. *From Eternity to Here: The Quest for the Ultimate Theory of Time*. Dutton, 2010.

6. "Causal Determinism." *Stanford Encyclopedia of Philosophy*, Stanford University, 23 Jan. 2003.

7. Feynman, Richard. "Nobel Lecture." Nobel Foundation, 11 Dec. 1965. Speech.

8. Frampton, P. H. *On Cyclic Universes.* 2006. *ArXiv,* arxiv.org/pdf/astro-ph/0612243v1.pdf. Accessed 17 Oct. 2016.

9. Gau, Shan. *Meaning of the Wavefunction. www.arxiv.org,* arxiv.org/pdf/1001.5085.pdf. Accessed 21 Oct. 2016.

10. Gibbons, G.W., and C.A.R. Heirdiro. *Supersymmetric Rotating Black Holes and Causality Violation. www.arxiv.org,* arxiv.org/abs/hep-th/9906098. Accessed 21 Oct. 2016.

11. Lawlor, Robert. *Voices Of The First Day: Awakening in the Aboriginal Dreamtime.* Rochester, Vermont, Inner Traditions International, 1991.

12. Lobo, Francisco. *Closed Timelike Curves and Causality. www.arxiv.org,* arxiv.org/abs/1008.1127. Accessed 21 Oct. 2016.

13. ---. *Time Machines and Traversable Wormholes in Modified Theories of Gravity. www.arxiv.org,* arxiv.org/abs/1212.1006. Accessed 21 Oct. 2016.

14. Mohsen, Razavy. *Quantum Theory of Tunneling.* World Scientific, 2003.

15. Nemiroff, Robert, and David Russel. *How Superluminal Motion Can Lead to Backward Time Travel.* *www.arxiv.org*, arxiv.org/abs/1505.07489. Accessed 21 Oct. 2016.

16. Penrose, Roger. "Before the Big Bang : An Outrageous New Perspective and its Implications for Particle Physics." *EPAC*, 2006, accelconf.web.cern.ch/accelconf/e06/PAPERS/THESPA01.PDF.

17. ---. *Cycles of Time: An Extraordinary View of the Universe.* Vintage, 2012.

18. Sean Carroll. *Spacetime and Geometry.* Pearson, 2004.

19. Shankar, R. *Principles of Quantum Mechanics.* Springer Science, 1984.

20. Sully, Marlan O. *A Delayed Choice Quantum Eraser.* 1999. *ArXiv*, arxiv.org/pdf/quant-ph/9903047.pdf.

A Path-Integral Formulation of Nonrelativistic Quantum Temporal Dynamics, and Implications on the Multiphase Discretization Metaverse Model

Abstract:

The present paper extends previous work developed by us in [4], [5], where a generalization of path-integration to a complex time-discretization index, suggesting the existence of a certain multiverse, was proposed, and time in nonrelativistic quantum mechanics was promoted from parameter to operator, resulting in a two-time model supporting Wheeler's well-known notion of a one-electron universe. In this article, we develop this quantum temporal dynamics from the path-integral formulation, reducing the foundations of the model to a few simple axioms. Using this formulation, we combine quantum temporal dynamics with the generalized path-integration, in the process providing a cosmological explanation for the location of the parallel universes developed in [5]. We comment on the effects upon temporal flow induced by the alternate time-discretization complex phases in quantum temporal dynamics, describing how superluminal travel and communication might be possible through the alternate universes of this model.

I. Introduction

A. Overview of the Multiphase Discretization Metaverse Model

In [5] the authors considered a generalization of the standard time-discretization framework of quantum mechanics, by generalizing the time-discretization index to arbitrary complex phase. As a result, the limit defining the value of the path-integral, given as

$$P(a,b) \cong \lim_{n \to \infty} \iiint \dots \int exp\left(\frac{i}{\hbar}S\right) dx_1 \dots dx_{n-1} \quad (1.1)$$

has a spectrum of different results when $n \in \mathbb{N} \to z \in \mathbb{C}$, depending on how z approaches infinity in the complex plane. Considering that (1.1), with complex index, represents the most general version of Feynman's path-integral, it is reasonable to suppose that these alternate values for the propagator $P(a,b)$, given by different directions of z in the complex plane, might correspond to different universes. The preceeding notion is developed along similar lines as Weinberg's generalization of the Einstein field equations, where a cosmological constant is introduced, effecting greater generality in the formulation of the equations, and different values of this constant are proposed to correspond to different universes. In a similar fashion, we take each different particle propagator, resulting from different limits of the complex time-slicing index z, as physical laws corresponding to distinct universes.

When this generalization is made, an integral must be computed over a complex number of distinct variables, and as such, a novel calculus was developed to determine the values of such expressions. In essence, the mathematical formalism of this theory follows similar lines as the fractional calculus, by generalizing certain properties of multiple integrals over an integer number of variables, to the case of complex dimension. For instance, the equation

$$\iint \dots \int \prod_{j=1}^{n} f(x_j) \, dx_1 \dots dx_n = \left(\int f(x) \, dx\right)^n \quad (1.2)$$

concerning the standard separation-of-variables property, is held to remain true even when $n \in \mathbb{N} \to z \in \mathbb{Z}$. Additional axioms of this calculus are also generalized from the integer case, including the substitution of variables, and the decomposition of functions into a Fourier integral. As a result, expressions of the form (1.1) can be interpreted even in the case of complex index, therefore allowing this generalization of path-integration to be effected.

In addition to generalizing the procedure of integration itself to the case of complex index, we also needed to generalize the path contribution formula, or integrand, to this new domain. As currently expressed, the Feynman contribution formula $exp\left(\frac{i}{\hbar}S\right)$ applies only to paths of integer discretization index, as this is the only domain where path-integration has been developed. Indeed, the formula $exp\left(\frac{i}{\hbar}S\right)$ is not purely motivated by the theoretical framework, and any number of other contribution formulae would be mathematically consistent and yield converging values for (1.1); this expression for the contribution formula is rather fixed by experiment, and this choice of path-integrand is necessary for the propagator expressed in (1.1) to be consistent with experiment, and produce the known form of the nonrelativistic Schrodinger equation. As such, there is no particular reason to expect that the contribution formula or integrand of generalized path-integration, where the index becomes complex, need be of the form $exp\left(\frac{i}{\hbar}S\right)$. In fact, any contribution formula F such that $F \to exp\left(\frac{i}{\hbar}S\right)$ where the time-slicing index $z \in \mathbb{N}$ is equally consistent in this regard, coinciding with the known formula in the region of integer time-slicing index where all experiments have been made.

Thus, there are many options available for generalizing the path contribution to the domain of complex time-discretization index. As such, we use primarily mathematical motivation to develop a reasonable generalization. We take as fundamental axioms the following postulates:

1. That the value of such expressions as (1.1), for complex n, be determined through generalizations of known properties of integer-dimensioned multiple integrals;
2. That the path-contribution formula is always a *pure phase factor* dependent upon the action.

This latter property itself forms the basis of Feynman's original path-integral framework, where Feynman's two postulates concern 1. That all paths contribute to the propagator, and 2. This contribution is given by a phase, dependent upon the action. It is indeed this phase property that results in the convergence upon the classical limit, as further explicated in Section II. As such, it appears reasonable to generalize these two postulates to complex index. Unfortunately, when the index is so generalized, the action S is no longer real, and the formula $exp\left(\frac{i}{\hbar}S\right)$ no longer a pure phase-factor. To amend this, we generalize the contribution rather as

$$exp(i\,\text{norm}(S)), \quad (1.3)$$

which reduces to the standard integrand for real time-slicing index, where S is real (the case of negative action is considered carefully in [5]).

We chose the p-norm for (1.3), as this general class of norms on $\mathbb{C}$ includes most useful norms on the complex numbers, including the Euclidean, taxicab, and maximum norms. As such, (1.3) reduces to a

TIME AND THE MULTIVERSE

formula of the form $exp\left(\frac{i}{\hbar}|S|_p\right)$. The action S is most reasonably defined as a (complex-order) sum, over each discretized leg of the path, of the integral of the Lagrangian over this leg. This definition reduces to

$$\sum_{j=0}^{n-1} \int_{t_j}^{t_{j+1}} \frac{m}{2}\left(\frac{(x_{j+1} - x_j)}{T} N\right)^2 - V\, dt \quad (1.4)$$

When such definitions for the path contribution are introduced to (1.1), and the new integration methods are thusly used, a spectrum of possible results are determined for the limit, corresponding to distinct particle propagators. These distinct results depend only upon the phase of the complex time-discretization index, and on the choice of p-norm to complete the generalization of the functional integrand. Although the value of this integrand, as in (1.3), is itself the exponential of a real number, the process of integration over a complex *number of* variables introduces non-trivial complex factors into the result. As [5] demonstrates, in the case of a free particle, this theory effectively introduces a complex factor multiplying the mass, given by

$$\frac{|\exp(it)|_p}{exp(it)} \quad (1.5)$$

which falls within a particular region of the complex plane, shown in Fig. 1.1.

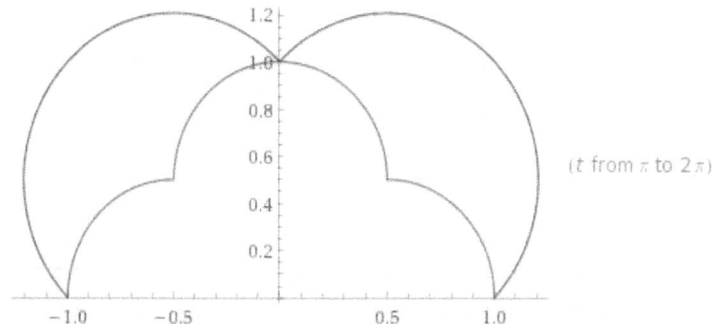

Fig. 1.1: The locations of the complex mass multipier in the complex plane are confined within the figure above.

When a potential is considered, a perturbation expansion must be used to determine the value of the propagator. This process is effected in [5], where a generalized Schrodinger equation is developed. This equation is given by the schema

$$-\frac{\hbar^2}{2m|\exp(i\mu)|_p}\frac{\partial^2 \psi}{\partial x^2} + \{|\cos(\mu)|^p - |\sin(\mu)|^p\}(|\cos(\mu)|^p + |\sin(\mu)|^p)^{\frac{1}{p}-1}V\psi$$

$$= \exp(-i\mu)\,i\hbar\frac{\partial \psi}{\partial t} \quad (1.6)$$

where the parameter μ, giving the discretization index phase, runs from π to 2π, while the norm parameter p is constrained to be greater than 1. Each different value of p and μ, corresponding to a point within the boundaries of Fig. 1.1, corresponds to a distinct version of the Schrodinger equation (1.6), interpreted as applying to a separate universe. In essence, each universe has a distinct effective mass multiplier (the point in Fig. 1.1), as well as a different factor $\{|\cos(\mu)|^p - |\sin(\mu)|^p\}(|\cos(\mu)|^p + |\sin(\mu)|^p)^{\frac{1}{p}-1}$ multiplying the potential.

B. Overview of Quantum Temporal Dynamics and the One-Electron Universe

In [4], the authors developed a novel theoretical framework to support Wheeler's notion of a one-electron universe. The latter idea developed from a consideration of the Feyman diagrams of quantum field theory, and particularly how antiparticles behave in the same fashion as particles moving in a reverse temporal direction. Wheeler's idea was that all electrons in the universe are, in fact, the *same* electron, the positron representing this particle reversing directions in time. The relativistic world-line of the electron therefore might appear to be a highly convoluted, "zig-zag" path, and the cross-section of space with this path contains separate electrons wherever this world-line intersects. The one-electron idea has, however, fallen our of favor, not least because it would require equal amounts of antimatter as physical matter. Various mechanisms of circumventing this, for example by postulating the existence of positrons "hiding" within the nucleus, have not met with experimental confirmation. Nonetheless, the one-electron universe model still serves as a useful pedagogical technique for introducing certain aspects of quantum field theory.

[4] developed a generalization of standard, nonrelativistic quantum mechanics, quantum temporal dynamics, that provides a possible mechanism for the one-electron universe. In usual nonrelativistic quantum theory, space is an operator while time is simply a parameter indexing the evolution of quantum states. As such, while wavefunctions and probability densities extend over the three spatial dimensions, they do not extend over time in the same fashion. Namely, one cannot speak of the probability of finding a particle in a certain temporal range, whatever that would mean, like one can speak of finding it in a certain spatial range. As such, while spatial kinematics and motion are consequences of the Schrodinger equation, the "motion" of particles through time goes unexplained.

Quantum temporal dynamics corrects this discrepancy, by describing temporal motion as a consequence of a more generalized Schrodinger equation. In order for QTD to work, two temporal dimensions must be considered, the time internal to a particle or system, and the time external to a particle or system. Internal time indexes the evolution of the particle's state, while external time indexes the interactions between particles and systems. Two systems can only interact if they coincide in external time. However, there is no such causal restriction on internal time, meaning that a particle can potentially interact with its past or future self.

Indeed, the latter notion forms the crux of QTD's one-electron universe. For simplicity, we consider a nonrelativistic universe consisting only of a single electron, with only an electrostatic force interaction given by the Coulomb potential, with instantaneous force transmission. Of course, our actual universe is far more complicated than this, but these results might be generalized to the relativistic framework of field theory. Furthermore, each different type of particle in our universe, for example quarks and muons, might correspond analogously to a single particle. The one electron in the one-electron universe of QTD reflects between two potential barriers located in the far future and distant past of external time, and the intersections of this world-line with any cross-section at a particular value of external time appear to us as separate particles.

Thus, the potential at any point of external time, and spatial position, is due entirely to this single particle. As the preceeding argument demonstrates, any internal-time state of the particle could conceivably contribute to this potential, so we should form an integral over all over internal time to determine the value of this potential. In [4], and in Section 2.1 of this paper, this formula is developed, given by

$$V(x,t) = R(t) + \iiint |\psi(r',t';T')|^2 V(r-r')\delta_\varepsilon(t-t')dr'dt'dT' \quad (1.7)$$

where t is external time and T is internal time. Here, the generalized Dirac Delta exponential $\delta_\varepsilon(x) = \frac{1}{\varepsilon\sqrt{2\pi}} exp\left(-\frac{x^2}{2\varepsilon^2}\right)$

determines the role of external time, by promoting contributions to the integral corresponding to states of the source particle close in external time. $R(t)$ describes the vast potential barriers in the future and past of external time. The function $V(\mathbf{r}-\mathbf{r}')$ is just the spatial Coulomb potential, given as $\frac{Dq^2}{|r-r'|}$ with q the charge on the electron and D some constant, whose value is determined in [4]. In that paper, we suggested that the value of ε, fixing the relative scale of violations of external time causality, could be roughly on the order of the Planck time. The reason we allowed such violations, and did not use a pure Dirac Delta function, was that the temporal wavefunctions are spread out over external time, and as such allowing this "fuzziness" seemed natural.

The second component of quantum temporal dynamics is the generalized Schrodinger equation, delineating the evolution of quantum temporal states in external time. We consider a wavefunction $\psi(\mathbf{r},t;T)$ over space and external time, evolving with internal time. The standard Schrodinger equation is obviously given as

$$-\frac{\hbar^2}{2m}\nabla^2\psi + V\psi = i\hbar\frac{\partial\psi}{\partial t} \quad (1.8)$$

Given that the d'Alembertian is simply the Laplacian on four-dimensional space-time, we thought it natural to generalize this equation to the temporal case as

$$-\frac{\hbar^2}{2m}\Box\psi + V\psi = i\hbar\frac{\partial\psi}{\partial T} \quad (1.9)$$

However, (1.9) must be modified to be made consistent with the true nature of the universe. A rescaling factor must be introduced before the temporal derivative, as the succeeding section of this paper and [4] make clear. Thus, we have

$$-\frac{\hbar^2}{2m}\Box_a\psi(r,t;T) + V\psi(r,t;T) = i\hbar\frac{\partial\psi(r,t;T)}{\partial T} \quad (1.10)$$

where the modified d'Alembertian $\Box_a = \nabla^2 - \frac{1}{a^2}\frac{\partial^2}{\partial t^2}$. The constant a is a free parameter of the theory, the "temporal velocity," whose nature is further explained in the succeeding section. We note that the negative sign before the temporal derivative above is reversed in the succeeding part of this paper, when quantum temporal dynamics is developed from the path-integral formulation instead.

Substituting (1.7), the generalized Schrodinger equation (1.10) yields

$$-\frac{\hbar^2}{2m}\Box_a\psi(r,t;T) + \Big\{R(t)$$
$$+ \iiint |\psi(r',t';T')|^2 V(r-r')\delta_\varepsilon(t-t')dr'dt'dT'\Big\}\psi(r,t;T)$$
$$= i\hbar\frac{\partial\psi(r,t;T)}{\partial T} \quad (1.11)$$

Assuming a simplified potential of the form $V(r) + R(t)$ that doesn't change over external time, and by introducing a separation of variables,

$$\psi(r,t;T) = \theta(r,T)\phi(t,T) \quad (1.12)$$

describing the quantum state as a product of spatial and temporal wavefunctions independently, we can decouple (1.11), producing

$$-\frac{\hbar^2}{2m}\nabla^2\theta(r,T) + V(r)\theta(r,T) = i\hbar\frac{\partial}{\partial T}\theta(r,T) \quad (1.13)$$

$$\frac{\hbar^2}{2ma^2}\frac{\partial^2}{\partial t^2}\phi(t,T) + R(t)\phi(t,T) = i\hbar\frac{\partial}{\partial T}\phi(t,T) \quad (1.14)$$

The form of (1.14) immediately lends credence to the notion of "temporal duality" developed in [4], namely that the temporal wavefunction ϕ behaves in exactly the same fashion as the standard spatial wavefunction, but rescaled by the temporal velocity coefficient a. As such, motion through external time can be understood in much the same way as motion through space. We imagine that the form of ϕ might be a very narrow wavepacket, reflecting between the great potential barriers defined by $R(t)$. It should also be noted that (1.13) is simply the standard Schrodinger equation, but with internal time in the place of standard time. As such, (1.13) demonstrates that it is internal time that evolves the quantum state of the particle, in accordance with our interpretation listed above.

C. Synthesis of QTD with Multiphase Discretization

The present paper firstly endeavours to develop Quantum Temporal Dynamics from the path-integral formulation, and thus motivating the generalized Schrodinger equation (1.11) from the basic axioms fixing the contribution formulae for paths through external time as well as space. Once such a path-integral formulation of QTD is developed, QTD can be applied in the case of a complex time-discretization index. As such, the temporal behavior of particles in the alternate universes induced by mutliphase discretization can be determined. As Section 5 demonstrates, according to the combination of QTD and Complex Discretization, particles in these parallel universes should travel slower through time than in our own universe, opening possibilities of superluminal travel and communication, described in Section 5.

Furthermore, the cosmology of QTD, developed in Section 4, suggests a possible physical location for the parallel universes of Section 1.1. As we argue below, these universes might exist in our future or past, separated by a quantum potential barrier. The single electron populates these universes by tunneling through these barriers. These chronologically separated domains might offer the location of the parallel dimensions explored previously.

II. Path-Integral Formulation of Quantum Temporal Dynamics

A. Fundamental Postulates of Temporal Dynamics

The framework of quantum temporal dynamics distinguishes between the *internal time* of a system, which indexes the change of its quantum state, and the *external time*, which indexes its relationship with other systems. As such, two systems can only interact if they coincide in external time; thus, a system can interact with its past state, or future state, if they coincide in external time, even if they differ significantly in internal time. In the one-electron universe, with large potential barriers at the past and future, the potential at a particular point of space-time is due entirely to a single particle, whose state evolves according to a nonlinear self-interaction as in (1.11).

In deriving this model of temporal dynamics from the functional-integral approach, given that internal time evolves the system while external time is an observable, it seems natural to treat external time, at least in the context of the free particle, in the same fashion as the spatial dimensions. The difference between external time, and a fourth spatial dimension, comes about in time's role as determining the interaction between systems, as in (1.7).

As such, let us consider the path-integral Lagrangian (1.4), but with multiple spatial dimensions:

$$\sum_{j=0}^{n-1}\int_{t_j}^{t_{j+1}}\frac{m}{2}\left(\frac{(x_{j+1}-x_j)^2+(y_{j+1}-y_j)^2+(z_{j+1}-z_j)^2}{T^2}N^2\right)-V\,dt \quad (2.1)$$

Introducing the external time parameter analogously, we have

$$\sum_{j=0}^{n-1}\int_{t_j}^{t_{j+1}}\frac{m}{2}\left(\frac{(x_{j+1}-x_j)^2+(y_{j+1}-y_j)^2+(z_{j+1}-z_j)^2+(\tau_{j+1}-\tau_j)^2}{T^2}N^2\right)$$

$-V\,dt \quad (2.2)$

where, to prevent confusion, we have relabelled external time as τ and internal time as t (where T is the total duration of internal time). However, as explicated in depth in [4], regarding the analogous Schrodinger formulation, this provisional formula is inconsistent with the known properties of the universe, and must be adequately modified. In essence, we can understand this as a consequence of time being more "rigid," in a sense, than space. That is, deviations from the standard classical trajectory through time are far smaller than the corresponding deviations in spatial motion. While in classical Lagrangian mechanics, the path of lowest action is always traversed, in the quantum path-integral formulation all paths contribute to the propagator; however, due to the phase factor $exp\left(\frac{i}{\hbar}S\right)$, paths with minimal action tend to reinforce their contributions, wheras paths with higher action have cancelling contributions. As S increases from the minimum action, changes in path correspond to smaller and smaller phase periods, thus leading to greater cancellation. Hence, in the classical limit, only the path of minimal action contributes to the propagator in a non-negligible way.

By "rigidity" of time, we mean that deviations of the path with respect to the external time parameter, from the classical path of least action, contribute to the propagator far less than corresponding deviations in the spatial coordinates. As such, let us multiply the external time parameter in (2.2) by some factor, to magnify the effect of changes of the path in external time:

$$\sum_{j=0}^{n-1} \int_{t_j}^{t_{j+1}} \frac{m}{2}\left(\frac{(r_{j+1}-r_j)^2 + a^2(\tau_{j+1}-\tau_j)^2}{T^2} N^2\right) - V \, dt \quad (2.3)$$

where henceforth we shall abbreviate $(x_i - x_{i-1})^2 + (y_i - y_{i-1})^2 + (z_i - z_{i-1})^2$ as $(r_i - r_{i-1})^2$. Here, a is a free parameter of the theory, the "temporal velocity." In [4], we argued that the value of a, to be consistent with experiment, must be roughly 10^{80} c. As Section IV will demonstrate, when combined with the multiphase discretization model, this must be amended to roughly 10^{100} c.

The addition of a might be considered arbitrary, as a substitution of variables could be introduced to subsume it, as in $\tau'_i - \tau'_{i-1} = a(\tau_i - \tau_{i-1})$. However, the nature of τ, as distinguished from any rescaled temporal parameter, is that it determines the same measurements as our standard units of time. That is, particle states evolve relative to t in such a manner that they obey the standard Schrodinger equation relative to this parameter, as in [1.13]. Thus, the specific parameter τ is such that the velocity of the particle through time, as given by the ratio of external time traversed to internal time, is nearly exactly 1. Hence, the two measures of time are consistent as they are traditionally used.

The equation (2.3), the Lagrangian for a path-integral temporal dynamics, is the first major postulate of the theory, expressing how a particle's location in external time evolves with internal time (by considering external time on the same footing as traditional spatial dimensions in the propagator). The second postulate concerns what was expressed at the end of 1.2, namely the nature of external time as defining the relationships between particles or systems. In our one-electron universe, we postulate that the potential at any point in space-time is due entirely to the influence of one particle, whose future and past states correspond to apparently different particles at any value of external time. Thus, all different states of the particle, corresponding to distinct values of internal time, can hypothetically contribute to this potential. Therefore, given the temporal wavefunction developed in (1.11), we have

$$V(\boldsymbol{r},\tau) = \iiint |\psi(\boldsymbol{r}',\tau';t')|^2 f(\boldsymbol{r}-\boldsymbol{r}',\tau-\tau')\,d\boldsymbol{r}'d\tau'dt' \quad (2.4)$$

where we integrate over all states of the particle at different internal times t to determine the value of the potential. Here, the contribution of the differential component of the temporal wavefunction ψ, at some location $(\boldsymbol{r}',\tau';t')$, is determined solely by its relative position in space and external time, given by the function $f(\boldsymbol{r}-\boldsymbol{r}',\tau-\tau')$. Generally, as τ indexes particle interations, significant differences $\tau-\tau' \gg 0$ diminish the value of the function f significantly. Only for $\tau-\tau' \approx 0$ is there a significant contribution to the potential.

In the case of electrostatic potential, which our simplified model assumes, we might ask why the form of f expressed in (1.7) is necessary; namely, why can't f simply adopt the standard form of the electrostatic potential, but with an additional spatial dimension, thus

$$f(\boldsymbol{r}-\boldsymbol{r}',\tau-\tau') = \frac{s}{\sqrt{(\boldsymbol{r}-\boldsymbol{r}')^2 + a^2(\tau-\tau')^2}} \quad (2.5)$$

for some constant s. If a is indeed sufficiently large, then $\tau-\tau' \gg 0$ will correspond to small values of f, consistent with our intended intepretation of external time. Futhermore, $\tau-\tau' = 0$ reduces to the standard potential formula, as given in Section 1.B. The difficulty here is that if we take values of internal time t' so that the temporal wavefunction is localized around an external time τ' that very slightly differs from τ, we can constrain $a^2(\tau - \tau')^2$ to have any arbitrary small but finite value. If the temporal wavefunction is not sufficiently thin to be so localized, then this problem also emerges, even when integrating the state at one value of internal time. As $\frac{s}{\sqrt{(r-r')^2+d}}$ for such a value d is not related to the standard potential formula by a simple factor, this choice of f is not consistent with the known properties of electrostatic potential. Thus, we are led rather to consider that form expressed in (1.7); namely,

$$V(r,\tau) = \iiint |\psi(r',\tau';t')|^2 V(r - r')\delta_\varepsilon(\tau - \tau')\, dr'd\tau'dt' \quad (2.6)$$

with an exponential Dirac Delta δ_ε of width ε. As explicated in [4], this formula produces the known experimental result for the potential due to a quantum wavefunction, the states of the single particle at different values of internal time t' apparently behaving as distinct particles for any "cross-section" of space-time at some external time τ. It must be again emphasized that this framework is non-relativistic, as the given Lagrangian is not relativistically invariant.

B. The Free Temporal Particle

Let us consider the behavior of our single particle, absent the form of the potential described in (2.6). As such, we apply the path contribution formula, determined by (1.4) and (2.1), to a free particle where the potential V vanishes, meaning that the electrostatic force considered in this model is assumed vanishing. A brief analysis of (1.4) suffices to demonstrate that the notion of temporal duality, explicated in Section 1.2 and [4], is justified from these path-integral axioms. Namely, the introduction of the external time-coordinate is equivalent to the introduction of a second spatial coordinate, but scaled by the "temporal velocity" constant a. Thus, acknowledging the separability of the corresponding Schrodinger equation as described in Section 1.2, we note that the motion of the particle through external time satisfies much the same principles as free quantum motion through a single spatial dimension, but modified by the $1/a$ term.

Alternatively, to derive the propagator directly, we follow the discretization program outlined in [4], rewriting (1.1) as

$$\lim_{n\to\infty} \iiint \cdots \int exp\left(A\sum_{j=0}^{n-1}\{x_{j+1}-x_j\}^2 + a^2(\tau_{j+1}-\tau_j)^2\right) dx_1 d\tau_1 \ldots dx_{n-1} d\tau_{n-1} \quad (2.7)$$

using the Lagrangian in (1.4), where $A = \frac{i}{\hbar}\frac{m}{2}\frac{n}{T}$. Here, T is the total interval of internal time, while we are simplifying by considering only a single spatial dimension, in addition to external time (these results trivially generalize to multiple spatial dimensions). First, we note that this expression readily separates to integrals individually over space and external time, as

$$\lim_{n\to\infty} \iiint \cdots \int exp\left(A\sum_{j=0}^{n-1}\{x_{j+1}-x_j\}^2\right) dx_1 \ldots dx_{n-1} \iiint \cdots \int exp\left(A\sum_{j=0}^{n-1} a^2(\tau_{j+1}-\tau_j)^2\right) d\tau_1 \ldots d\tau_{n-1} \quad (2.8)$$

Performing a standard variable substitution, we have

$$\lim_{n\to\infty} \iiint \cdots \int exp\left(A\left\{\sum_{j=1}^{n-1}\{z_j\}^2 + \left(x_n - x_0 - \sum_{i=1}^{n-1} z_i\right)^2\right\}\right) dz_1 \ldots dz_{n-1} \iiint \cdots \int exp\left(Aa^2\left\{\sum_{j=1}^{n-1}\{w_j\}^2 + \left(\tau_n - \tau_0 - \sum_{i=1}^{n-1} w_i\right)^2\right\}\right) dw_1 \ldots dw_{n-1} \quad (2.9)$$

where $z_j = x_j - x_{j-1}$ and similarly for w_j. For simplicity's sake, we now simplify just the latter integral, the former following trivially.

Introducing a Fourier integral expansion, we have

$$\lim_{n \to \infty} \int_{-\infty}^{\infty} exp\left(\frac{k^2}{4a^2 A}\right) \iiint \dots \int exp\left(a^2 A \sum_{j=1}^{n-1} \{w_j\}^2\right) exp\left(ik\left(\tau_n - \tau_0\right.\right.$$
$$\left.\left. - \sum_{i=1}^{n-1} w_i\right)\right) dz_1 \dots dz_{n-1}\, dk \quad (2.10)$$

Applying the variable separation of (1.2), we have, simplifying this integral,

$$\lim_{n \to \infty} \int_{-\infty}^{\infty} exp\left(\frac{k^2}{4a^2 A}\right) exp(ik(\tau_n - \tau_0)) \left\{\int exp(a^2 A z^2 - ikz) dz\right\}^{n-1} dk \quad (2.11)$$

yielding

$$exp\left(\frac{ma^2 \tau^2 i}{2\hbar T}\right) \quad (2.12)$$

where τ is the total duration of external time traversed by the particle. Indeed, in accordance with the notion of temporal duality, we see that this formula is equal to the propagator for the standard free particle in quantum mechanics, but with the spatial component rescaled by the temporal velocity factor. Applying the same procedure to the spatial integral, we have

$$exp\left(\frac{mX^2 i}{2\hbar T}\right) exp\left(\frac{ma^2 \tau^2 i}{2\hbar T}\right) \quad (2.13)$$

which is again consistent with our framework of temporal duality, and with the generalized Schrodinger equation expressed in (1.11).

C. Deriving the Schrodinger-Equation for the One-Electron Universe

Applying the path-integral framework to the particle in a potential is a relatively straightforward application of the perturbative method, as expressed in [5]. In quantum temporal dynamics, where the universe is modelled as a one-electron system, the potential at any point in space-time is due entirely to this single particle, given as an integral of contributions over the particle's entire internal time history, as expressed in (2.6). Incorporating the large potential barriers at the being and end of the universe, that induce the reflections of the single particle, as described in Section 1, we have as our potential

$$V(\boldsymbol{r},\tau) = R(\tau) + \iiint |\psi(\boldsymbol{r}',\tau';t')|^2 V(\boldsymbol{r} - \boldsymbol{r}')\delta_\varepsilon(\tau - \tau')\, d\boldsymbol{r}'d\tau'dt' \quad (2.14)$$

where $R(\tau)$ constitutes these large potential barriers. Assuming the function ψ is given, this expression defines a function of $(\boldsymbol{r}, \tau)$, the spatial and external time coordinate. In the simplification of a single spatial dimension x, we have

$$V(x,\tau) = R(\tau) + \iiint |\psi(x',\tau';t')|^2 V(x - x')\delta_\varepsilon(\tau - \tau')\, dx'd\tau'dt' \quad (2.15)$$

where, if ψ is known, $V(x, \tau)$ is again a well-defined function of space and external time. Using the Lagrangian defined in (2.3), we have that the functional integrand, the contribution corresponding to each path, is given as

$$exp\left(\frac{i}{\hbar}\left(\left[\sum_{j=0}^{n-1}\frac{mn}{2}\frac{\{x_{j+1}-x_j\}^2+a^2(\tau_{j+1}-\tau_j)^2}{T}\right]\right.\right.$$

$$-\frac{T}{n}\sum_{j=0}^{n-1}\int_0^1 V(x_j+(x_{j+1}-x_j)s,\tau_j$$

$$\left.\left.+(\tau_{j+1}-\tau_j)s)ds\right)\right) \quad (2.16)$$

Incorporating this contribution formula into the functional integral, the expression in (1.1), we have

$$\lim_{n\to\infty}\left(K-\frac{i}{\hbar}\frac{1}{n}\int\int\cdots\int exp\left(\frac{i}{\hbar}S[0]\right)\left(T\sum_{j=0}^{n-1}\int_0^1 V(x_j+(x_{j+1}-x_j)s,\tau_j\right.\right.$$

$$\left.\left.+(\tau_{j+1}-\tau_j)s)ds\right)dx_1 d\tau_1\ldots dx_{n-1}d\tau_{n-1}+\cdots\right) \quad (2.17)$$

where $S[0]$ is the free-particle action given by (2.16) with $V = 0$, and K is the free-particle propagator given in (2.13). Now, when deriving the Schrodinger equation, we will assume an infinitesimal internal time-translation T, as Feynman does in [2] and we perform in [5]. This simplification allows us to transform (2.17) into

$$\lim_{n\to\infty}K-\frac{i}{\hbar}\frac{1}{n}\int\int\cdots\int exp\left(\frac{i}{\hbar}S[0]\right)\left[exp\left(T\sum_{j=0}^{n-1}\int_0^1 V(x_j\right.\right.$$

$$\left.+(x_{j+1}-x_j)s,\tau_j+(\tau_{j+1}-\tau_j)s)ds\right)$$

$$-1\bigg]dx_1 d\tau_1\ldots dx_{n-1}d\tau_{n-1}+\cdots \quad (2.18)$$

where the potential term is written in exponential form, assuming again the infinitesimal nature of the internal time interval. Now, as we are assuming the total internal time-interval to be infinitesimal, we can further stipulate that $x_j + (x_{j+1} - x_j)s \to x_a$, as the argument of the potential in the integral is now dominated by the path's initial starting point rather than the relatively infinitesimal displacement, and likewise for external time, allowing us to write

$$\lim_{n\to\infty} K - \frac{i}{\hbar}\frac{1}{n} \int \int \cdots \int exp\left(\frac{i}{\hbar}S[0]\right)\left[exp\left(T\sum_{j=0}^{n-1} V(x_a, \tau_a)\right) - 1\right] dx_1 \ldots dx_{n-1} + \cdots \quad (2.19)$$

The straightfoward substitution and separation of variables, and Fourier integral expansion, as applied in the previous sub-section, but now with the appearance of the sum $\sum_{j=0}^{n-1} T\, V(x_{a_j}, \tau_a)$, yields

$$\lim_{n\to\infty} K - \frac{i}{\hbar}\frac{1}{n}\left(K exp(nTV(x_a, \tau_a)) - K\right) + \cdots \quad (2.20)$$

Which, for small T, gives

$$\lim_{n\to\infty} K - \frac{i}{\hbar}\left(KTV(x_a, \tau_a)\right) + \cdots \quad (2.21)$$

Producing[9]

$$K - \frac{i}{\hbar}TV(x_a, \tau_a)K + \cdots \quad (2.22)$$

Applying the definition of the wavefunction in terms of the propagator, as we do in [5] and Section 3.3 below, results in the equation

$$-\frac{\hbar^2}{2m}\Box_a \psi(x, \tau; t) + V(x, \tau)\psi(r, \tau; t) = i\hbar\frac{\partial \psi(x, \tau; t)}{\partial t} \quad (2.23)$$

---

[9] [5] provides a full derivation of the analogous formula for complex time-discretization index.

where $\Box_a = \nabla^2 + \frac{1}{a^2}\frac{\partial^2}{\partial t^2}$. In the single spatial dimension we are considering, the Laplacian ∇^2 readily reduces to the partial derivative $\frac{\partial^2 \psi(x,\tau;t)}{\partial x^2}$. This equation is simply the standard Schrodinger equation in two spatial dimensions, but with the second dimension (delineating external time) rescaled, and the potential $V(x,\tau)$ given by the formula (2.15). Application of separation of variables into wavefunctions over space and external time separately readily results in (1.13-14), which defines the duality of temporal and spatial motion.

D. Differences with the Quantum Field Theory Promotion of Time to an Operator

The notion of promoting time to an operator, rather than demoting space to a parameter, has been explored before in the context of quantum field theory, for example in [6]. As such, a Schrodinger equation results, defining the dynamics of the particle's motion through time. In order to work, this framework requires two separate temporal dimensions, the operator considered above, and a temporal paramater to evolve quantum states. The latter is the relativistic proper time of the particle, while the former is the time as measured in some reference-frame external to the particle.

We note that our theory, which also describes two temporal components (internal and external to the particle), one of which is promoted to operator, is unique and not in any way similar to the quantum field theory notion. Firstly, quantum temporal dynamics is a nonrelativistic theory, and therefore separate from the mechanisms of quantum field theory. The motivation to promote time to operator is not to ensure relativistic invariance or space-time symmetry; indeed, the form of the Lagrangian in (2.3) is not relativistically invariant. As such, there is no room for the relativistic notion of "proper time" in this theory. The difference between internal and external time is completely separate from the relationship between proper time, and time as measured by some external observer. The latter two notions are linked by relativistic kinematics, meaning that if the particle is moving quickly relative to that observer, then its speed through external time, measured relative to its proper time, will greatly increase, while there is no such relationship in quantum temporal dynamics. The speed of a particle in QTD through external time, as measured by the ratio of external time traversed to internal time elapsed, bears no relationship to the spatial velocity of the particle in some reference-frame, as the temporal duality expressed in [1.14] makes perfectly obvious.

Furthermore, the notion of promoting time to operator has never before been linked to the one-electron universe hypothesis. As the previous Section describes, the notion of quantum temporal dynamics was developed specifically to support a dynamical version of the one-electron universe. As such, the potential at any point in space-time is given solely by the single particle, by an integral over its entire future and past history. External time coordinates serve, through a generalized Dirac Delta function, to determine the relative contribution of each internal time state of the particle to this potential, while large potential barriers are present at the beginning and end of the universe. These notions are entirely absent from the QFT formulations involving proper time as parameter and some external time coordinate as operator. Namely, the self-interaction formula (2.15) and Lagrangian (2.3), yielding the temporal Schrodinger equation (2.23), do not appear in any previous work.

III. Quantum Temporal Dynamics with Complex Discretization Index

A. Solving the Free-Particle of Temporal Dynamics for Complex Discretization

The purpose of the present paper is to combine the mathematical formalisms of quantum temporal dynamics with that of path-integration for complex discretization index. As such, we shall apply the results of the previous section to the case of complex slicing index, and derived a generalized form of the Schrodinger equation that specifies the temporal dynamics given a discretization index of complex phase. In essence, this shall be a temporal version of (1.6) specifying the dynamics of particles in external time, in the different universes specified by alternate phases of the index. First, we follow the general outline of Section 2.2, and derive the behavior of the free temporal particle, without the presence of the self-interacting potential (2.15).

Firstly, we have the contribution formula (1.3), given formula (2.3) for the Lagrangian, reduce to

$$exp\left(\frac{i}{\hbar}\left|\sum_{j=0}^{z-1}\int_{t_j}^{t_{j+1}}\left(\{x_{j+1}-x_j\}^2+a^2(\tau_{j+1}-\tau_j)^2\right)\frac{z^2}{T^2}dt\right|_p\right) \quad (3.1)$$

In [5], in developing the mathematical mechanism of such functional integration, we assume that the real or imaginary part of a sum is the sum of the real or imaginary parts of the terms,

$$Re\left(\sum_{j=0}^{z-1}\int_{t_j}^{t_{j+1}}\left(\{x_{j+1}-x_j\}^2+a^2(\tau_{j+1}-\tau_j)^2\right)z^2\,dt\right)$$

$$=Re(z)\left[\sum_{j=0}^{z-1}T\left(\{x_{j+1}-x_j\}^2+a^2(\tau_{j+1}-\tau_j)^2\right)\right] \quad (3.2)$$

Likewise,

$$Im\left(\sum_{j=0}^{z-1}\int_{t_j}^{t_{j+1}}\left(\{x_{j+1}-x_j\}^2+a^2(\tau_{j+1}-\tau_j)^2\right)z^2\,dt\right)$$

$$=Im(z)\left[\sum_{j=0}^{z-1}T\left(\{x_{j+1}-x_j\}^2+a^2(\tau_{j+1}-\tau_j)^2\right)\right] \quad (3.3)$$

Implying that

$$\left|\sum_{j=0}^{z-1}\int_{t_j}^{t_{j+1}}\left(\{x_{j+1}-x_j\}^2+a^2(\tau_{j+1}-\tau_j)^2\right)z^2\,dt\right|_p$$

$$=|n|_p\left[\sum_{j=0}^{z-1}T\left(\{x_{j+1}-x_j\}^2+a^2(\tau_{j+1}-\tau_j)^2\right)\right] \quad (3.4)$$

Hence, the functional integral (1.1) readily reduces to

$$\lim_{z\to\infty}\iiint\cdots\int\exp\left(A\sum_{j=0}^{z-1}\{x_{j+1}-x_j\}^2\right.$$

$$\left.+a^2(\tau_{j+1}-\tau_j)^2\right)dx_1d\tau_1\ldots dx_{z-1}d\tau_{z-1} \quad (3.5)$$

where

$$A=\frac{i}{\hbar}\frac{m}{2}\frac{|n|_p}{T} \quad (3.6)$$

Now, to compute this expression, we must introduce an additional mathematical postulate to the mechanism expressed in [5], namely that an alternating multiple integral of complex dimension, of the form

$$\iiint \cdots \int f\left(\sum_{j=0}^{z-1} g(x_{j+1} - x_j)\right) w\left(\sum_{j=0}^{z-1} h(\tau_{j+1} - \tau_j)\right) dx_1 d\tau_1 \ldots dx_{z-1} d\tau_{z-1} \quad (3.7)$$

for analytic and explicitly defined functions f, g, w, h, can be readily decomposed as

$$\iiint \cdots \int f\left(\sum_{j=0}^{z-1} g(x_{j+1} - x_j)\right) w\left(\sum_{j=0}^{z-1} h(\tau_{j+1} - \tau_j)\right) dx_1 d\tau_1 \ldots dx_{z-1} d\tau_{z-1}$$

$$= \iiint \cdots \int f\left(\sum_{j=0}^{z-1} g(x_{j+1} - x_j)\right) x_1 \ldots dx_{z-1} \iiint \cdots \int w\left(\sum_{j=0}^{z-1} h(\tau_{j+1} - \tau_j)\right) d\tau_1 \ldots d\tau_{z-1} \quad (3.8)$$

As such, (3.8) being a postulate for complex-dimensioned integration could be viewed as a similar separation axiom to (1.2), but performing a double, rather than complex-order, separation. Applying (3.8) to (3.5), we have

$$\lim_{z \to \infty} \iiint \cdots \int \exp\left(A \sum_{j=0}^{z-1} \{x_{j+1} - x_j\}^2\right) dx_1 \ldots dx_{z-1} \iiint \cdots \int \exp\left(A \sum_{j=0}^{z-1} a^2(\tau_{j+1} - \tau_j)^2\right) d\tau_1 \ldots d\tau_{z-1} \quad (3.9)$$

The derivation from this point is straightforward, following the mathematical route proscribed in Section 2.2, and in the main body of [5]. Each factor simplifies in the same manner as the standard free-particle for complex discretization index, described in [5], by an application of substitution of variables to separate the exponential factors within the integrand, the use of a Fourier integral expansion to achieve this, and use of (1.2). Thus, the result is simply the multiplication of the two standard propagators produced by each factor of (3.9), namely

$$exp\left(\frac{mX^2 i\left(|\exp(i\mu)|_p\right)}{2\hbar T exp(i\mu)}\right) exp\left(\frac{ma^2\tau^2 i\left(|\exp(i\mu)|_p\right)}{2\hbar T exp(i\mu)}\right) \quad (3.10)$$

μ being the complex phase of the time-discretization index, running from π to 2π, p again determining the choice of norm, and a being the temporal velocity. (3.10) affirms the notion of temporal duality appearing in Section 2.2, explicated in Section 1.2 and [4]. As such, the quantum propagator corresponding to the motion through external time is merely a rescaled form of that corresponding to spatial motion, as the form of the Lagrangian (2.3) makes evident.

B. The Particle in a Potential with Complex Discretization

The determination of quantum temporal dynamics, for complex time-discretization index, in a potential follows straightforwardly from the programme adapted in Section 2.3. Firstly, however, we must note that in the other universes developed in [5], standard unitarity does not apply to the quantum wavefunction, meaning that the wavefunctions solving (1.6) are generally not normalizable. While the authors of [5] did develop a probabilistic interpretation to account for such non-normalizability, unfortunately this interpretation is inconsistent with the cosmological model set forth in Section 4. As such, we shall define the meaning of the wavefunction such that

$$p(x', \tau'; t') = A(t)|\psi(x', \tau'; t')|^2 \quad (3.11)$$

modifying the standard formula for the probability density p with the appropriate normalization factor $A(t)$ over internal time. The wavefunction itself is still given by the standard Feynman integral of the initial wavefunction and particle propagator, as expounded in [2]. As such, for complex discretization index, it appears that the appropriate definition for the potential at a point modifies (2.15) as

$$V(x,\tau) = R(\tau) + \iiint \frac{|\psi(x',\tau';t')|^2}{\int_{x'-\tau'} |\psi(x',\tau';t')|^2 dx'd\tau'} V(x-x') \delta_\varepsilon(\tau-\tau') \, dx'd\tau'dt' \quad (3.12)$$

Again, if the wavefunction ψ is a known function, (3.12) defines a perfectly well-defined function on (x,τ). As such, our strategy shall be to "assume" the wavefunction known, and thus (3.12) a defined function on space-time, carry (3.12) along with the derivation of the generalized Schrodinger equation, and ultimately define a nonlinear mixed integral-differential equation for the wavefunction.

Following the general procedure used in [5], we set the value of the path contribution as

$$exp\left(\frac{i}{\hbar}\left|\int_0^T L\,dt\right|_p\right) =$$

$$exp\left(\frac{i}{\hbar}\left(\left|Re(z)\left[\sum_{j=0}^{z-1}\frac{m}{2}\frac{\{x_{j+1}-x_j\}^2 + a^2(\tau_{j+1}-\tau_j)^2}{T}\right.\right.\right.\right.$$

$$\left.-TRe\left[\frac{1}{z}\right]\sum_{j=0}^{z-1}\int_0^1 V(x_j + (x_{j+1}-x_j)s, \tau_j + (\tau_{j+1}-\tau_j)s)ds\right|^p$$

$$+\left|Im(z)\left[\sum_{j=0}^{z-1}\frac{m}{2}\frac{\{x_{j+1}-x_j\}^2 + a^2(\tau_{j+1}-\tau_j)^2}{T}\right.\right.$$

$$-TIm\left[\frac{1}{z}\right]\sum_{j=0}^{z-1}\int_0^1 V(x_j + (x_{j+1}-x_j)s, \tau_j$$

$$\left.\left.\left.+ (\tau_{j+1}-\tau_j)s)ds\right|^p\right)^{1/p}\right) \quad (3.13)$$

Inserting this formula into (1.1) in the case of non-zero potential V, expanding the contribution formula in a Taylor series in

$-\sum_{j=0}^{n-1} \int_0^1 V(x_j + (x_{j+1} - x_j)s, \tau_j + (\tau_{j+1} - \tau_j))ds$, and simplifying, we obtain

$$\lim_{n\to\infty} \left(K - \frac{i}{\hbar} B \int \int \dots \int \exp\left(\frac{i}{\hbar} S[0]\right) \left(T \sum_{j=0}^{n-1} \int_0^1 V(x_j + (x_{j+1} - x_j)s, \tau_j \right.\right.$$

$$\left.\left. + (\tau_{j+1} - \tau_j)s)ds \right) dx_1 d\tau_1 \dots dx_{z-1} d\tau_{z-1} + \dots \right) \quad (3.14)$$

In which the ellipsis denotes the sum of the other terms, S[0] is the free particle action, and B is

$$B = \left(Re(n) Re\left(\frac{1}{n}\right) |Re(n)|^{p-2} + Im(n) Im\left(\frac{1}{n}\right) |Im(n)|^{p-2} \right) (|Re(n)|^p$$

$$+ |Im(n)|^p)^{\frac{1}{p}-1} \quad (3.15)$$

Again, to derive an analogue of the Schrodinger equation, as in 2.3, we need only consider infinitesimal time-translations T. For the limit of small T, as in 2.3, we have

$$\lim_{n\to\infty} K - \frac{i}{\hbar} B \int \int \dots \int \exp\left(\frac{i}{\hbar} S[0]\right) \left[\exp\left(T \sum_{j=0}^{z-1} \int_0^1 V(x_j \right.\right.$$

$$\left.\left. + (x_{j+1} - x_j)s, \tau_j + (\tau_{j+1} - \tau_j)s)ds \right)\right.$$

$$\left. - 1 \right] dx_1 d\tau_1 \dots dx_{z-1} d\tau_{z-1} + \dots \quad (3.16)$$

where the substitution of the exponential is accurate for small T. Furthermore, for such small time-translations, we may make the further simplification, as used in Section 2.3 and [5],

$$\lim_{n\to\infty} K - \frac{i}{\hbar} B \int \int \cdots \int \exp\left(\frac{i}{\hbar} S[0]\right) \left[\exp\left(T \sum_{j=0}^{z-1} V(x_a, \tau_a)\right)\right.$$

$$\left. - 1\right] dx_1 d\tau_1 \ldots dx_{z-1} d\tau_{z-1} + \cdots \quad (3.17)$$

as the value of the potential V is nearly exactly $V(x_a, \tau_a)$, over the entire path, for small T. Following the general procedure of Section 2, again with the appearance of the sum $\sum_{j=0}^{n-1} T\, V(x_a)$, we have

$$\lim_{n\to\infty} K - \frac{i}{\hbar} B(K\exp(nTV) - K) + \cdots \quad (3.18)$$

Which, for small T, yields

$$\lim_{n\to\infty} K - \frac{i}{\hbar} Bn(KTV) + \cdots \quad (3.19)$$

Producing

$$K - \frac{i}{\hbar} DTVK + \cdots \quad (3.20)$$

where V denotes the nearly uniform value of the potential along the infinitesimal path (equal to $V(x_a, \tau_a)$ above). In this case,

$$D = \exp(it)\{|\cos(t)|^p - |\sin(t)|^p\}(|\cos(t)|^p + |\sin(t)|^p)^{\frac{1}{p}-1} \quad (3.21)$$

Given the formula (3.10). Indeed, this is simply the standard formula for complex discretization index obtained in [5], but with the new free-particle propagator and potential, once again lending credence to the general principle of temporal duality. This is straightforward, as this demonstration was little more than an application of complex time-discretization to two spatial dimensions, one of them rescaled, and with the particular potential formula given by (3.12).

C. The Generalized Temporal Schrodinger Equation

Using the previously explicated definition of the wavefunction's evolution in terms of the particle propagator, and keeping in mind that the actual probability density is defined by a normalized function as in (3.11), we can now determine the generalized Shrodinger equation. We follow the same programme delineated in [5].

Relevant to the construction of the differential equation of the wavefunction is the behavior of the wavefunction over infinitesimal time-translations, and therefore the propagator over small T. Thus, let us consider the case of an arbitrarily small time-interval $T = \xi$. In this case, the approximations of 3.2 are accurate, and the propagator is given by (10),

$$K - \frac{i}{\hbar}D\xi VK + O(\xi^2) \quad (3.22)$$

or,

$$K exp\left(-\frac{i}{\hbar}D\xi V(x_a, \tau_a)\right) \quad (3.23)$$

Use of variable substitution, power series expansion, and the definition of the wavefunction in terms of the Kernel, as applied similarly in [5], suffices to demonstrate that

$$-\frac{\hbar^2}{2mz}\Box_a\psi + DV\psi = i\hbar\frac{\partial\psi}{\partial t} \quad (3.24)$$

with z given as in (1.5) and D as in (3.21).

Explicitly, the generalized Schrodinger equation is given as

$$-\frac{\hbar^2}{2m|\exp(i\mu)|_p}\square_a\psi(x,\tau;t) + \{|\cos(\mu)|^p$$

$$- |\sin(\mu)|^p\}(|\cos(\mu)|^p + |\sin(\mu)|^p)^{\frac{1}{p}-1}\bigg(R(\tau)$$

$$+ \iiint \frac{|\psi(x',\tau';t')|^2}{\int_{x'-\tau'}|\psi(x',\tau';t')|^2 dx'd\tau'} V(x-x')\delta_\varepsilon(\tau$$

$$-\tau')\,dx'd\tau'dt'\bigg)\psi(x,\tau;t)$$

$$= \exp(-i\mu)\,i\hbar\frac{\partial\psi(x,\tau;t)}{\partial t} \quad (3.25)$$

with the parameter μ running from π to 2π. (3.25) is the central equation describing quantum temporal dynamics in a complex time-discretized universe. As such, it reflects the temporal behavior of particles within the alternate realities described in Section 1.1. The appearance of temporal duality is apparent, in the sense that the equation is readily decomposed into two separate equations, assuming certain approximations of the potential, describing spatial and temporal motion respectively, that are rescaled versions of each other. (3.25) can be so separated using the techniques applied in [4]. At this point, it is interesting to consider how the quantum temporal one-electron cosmology relates to the complex time-discretized multiverse.

IV. Cosmological Multiverse Interpretation

A. Introduction and Formula for the Temporal Velocity (Temporal Stability Coefficient)

In [4], we presented a multiverse cosmology arising from the framework of temporal quantum dynamics. Namely, given the temporal duality expressed in (1.14), the interactions of the model's single electron with the potential barriers $R(\tau)$, from (2.14), should be equivalent to the interactions of a standard particle in one spatial dimension with such a barrier. As such, each reflection from the barrier also carries a small but nonzero chance of quantum tunneling beyond the barrier. If the particle does indeed tunnel in this fashion, beyond the range of external time constituting our "universe," then it will have entered an alternate domain, that we could thusly fashion as constituting an alternate "universe." Indeed, there might be an infinite chain or sequence of such potential barriers, the space between any two barriers constituting a separate universe. When the single electron tunnels through a barrier, it populates the next universe.

This cosmological model affords an interesting opportunity to introduce the multiverse of multiphase discretization. Namely, there is no particular reason to expect that the laws of physics applying in these alternate domains, in our far past and future and separated by vast energy barriers, are the same as in our universe. Indeed, the physical laws applying in these alternate realities could well be very different, as long as they are consistent with the general path-integral framework that forms the basis of quantum temporal dynamics. As such, there is no particular reason to expect that the complex phase of the path-discretization index, which is 0 in our universe, might not adopt a number of different values in these alternate domains. In fact, the multiverse naturally produced by quantum temporal dynamics offers an interesting model for the physical instantiation of the parallel universes of MDM, as well as a physical mechanism (quantum tunneling through the temporal energy barrier) by which these universes are populated. Thus, let us consider such a multiverse.

Firstly, we shall use a physical argument to determine the approximate value of the temporal velocity (or stability coefficient) a. Noting the nature of temporal duality, the temporal wavefunction (as in (1.14)) of the electron should behave in an analogous fashion to the standard quantum wavefunction of a particle in a single spatial dimension. We imagine that such temporal motion is defined by a very tight "wave packet," with the extent of the wavefunction in external time confined roughly to the Planck time. Taking the standard formula for the wavepacket of a particle, and multiplying the mass by the appropriate coefficient to effect the duality, we have (in atomic units)

$$\phi(\tau,t) = \left(\frac{2c}{\pi}\right)^{1/4} \frac{\exp\left(\frac{-c\tau^2 + i(l\tau - l^2 t/2ma^2)}{1 + 2ict/ma^2}\right)}{\sqrt{1 + 2ict/ma^2}} \quad (4.1)$$

where l is defined as the initial momentum of this wavepacket formation (in some sense, a "temporal momentum"), such that

$$\frac{l}{ma^2} = v \quad (4.2)$$

v being the velocity of the particle through external time, while c is a free parameter determining its width in external time. This solution, of course, applies only for real path-discretization, which we shall be using in this sub-section. The use of complex path-discretization, which we shall be considering in the succeeding sub-section and in Section 5, does not significantly alter these results.

Let us suppose that the total duration of internal time traversed by the particle, as it populates the entire multiverse, is T. In the succeeding sub-section we shall develop an approximate value for T. For now, we note that, as in [4], (4.1) gives the probability density (in atomic units)

$$|\phi(\tau,t)|^2 = \sqrt{2/\pi}\, w\, exp(-2w^2(\tau - vt)^2) \quad (4.3)$$

where

$$w = \left(\frac{c}{1 + (2ct/ma^2)^2}\right)^{1/2} \quad (4.4)$$

Applying the same procedure as in [4], we note that the width of the probability density Gaussian is proportional to

$$\frac{1}{w}$$

Thus, the width after an internal time T is given by

$$\left(\frac{1 + (2(1/d)^2 T/ma^2)^2}{(1/d)^2}\right)^{1/2} \quad (4.5)$$

where d is the initial width at internal time $T = 0$. To ensure consistency, the temporal wavepacket should preserve its Planck-scale width over its entire journey. If there are N particles within a particular universe in the QTD multiverse, we can infer that the electron makes N laps between the two potential reflectors $R(t)$ defining the extent of that universe in external time. Assuming that there are M separate universes, and that each universe lasts for an external time Y, the total internal time experienced by our single particle should be on the order of NMY. Thus, the dispersion of the temporal wavepacket over this journey is proportional to

$$\frac{NMY}{mda^2} \quad (4.6)$$

with d the initial width, and the internal time NMY given in atomic units. As the Planck time is 10^{-26} in atomic units, and the electron's mass is 1, our requirement reduces to

$$\frac{NMY}{(a^2 10^{-26})} \approx 10^{-26} \quad (4.7)$$

giving the "effective temporal mass"

$$a^2 = NMY10^{52} \quad (4.8)$$

in atomic units.

B. The Anthropic Principle and Universe Number

Consideration of the anthropic principle, as presented in [5], allows us to determine the approximate number of universes N in the combined MDM-QTD multiverse. Firstly, we must make an assumption concerning the distribution of physical parameters and properties to these distinct universes. Let us therefore assume that the quantum phase parameter μ in (3.25) is distributed evenly among the universes considered in the previous sub-section. If we discover an anthropic constraint concerning the value of μ in our universe, such that it lies in a range of length $1/L$, we can therefore estimate that the number of universes is at least on the order of L to accommodate this anthropic constraint without violation of reasonable probabilistic considerations.

Our anthropic constraint derives from a particular property of multiphase discretization, namely that, even in the presence of time-dependent normalization as in (3.11), the law of energy and momentum conservation is violated to varying degrees, depending on the quantum discretization phase μ corresponding to the relevant universe. Let us consider the behavior of quarks within nucelons, and stipulate that the structure of nucleons should be stable, even over long periods of time, within our universe. We can estimate that the time Y, between the hypothetical singularities at the beginning and end of our universe, is on the order of roughly 10^{10} years. Using a nonrelativistic analysis of the quark, we know that, confined within such a small location as a nucleon, by the Heisenberg uncertainty principle it has a proportionally large uncertainty in momentum. Its wavefunction in momentum space should therefore be of the form

$$exp(-p^2/a) \quad (4.9)$$

with a on the order of the quark's momentum uncertainty. Thus, we have the equivalent equation

$$exp(-E/r) \quad (4.10)$$

for the momentum-space equation, where E is kinetic energy and r is the "average" energy from the quantum uncertainty in momentum. It is readily determined from the uncertainty principle that, given the scale of the nucleon's radius, the value of r is roughly 10^9 hartrees. From the spatial Schrodinger equation of the form (1.13) arising from the general equation (3.25), we have that the time evolution of (4.10) is given as

$$exp(-E/r)\,exp(sin(\mu)Et) \quad (4.11)$$

in atomic units, with $t \approx \tau$ the particle's internal time, roughly equal to external time when the velocity v is roughly equal to 1. Thus, if we wish the average energy of the quarks to maintain the same order of magnitude, meaning that their uncertainty in momentum and hence in position also maintain the same order of magnitude, we should have

$$\mu t \approx \frac{1}{r} \quad (4.12)$$

in atomic units, giving us that μ is roughly 10^{-43}. We do not perform this calculation for the whole of the particle's internal-time, just for one "leg" between the temporal barriers, as the nucleons disintegrate and reform at the extremely high-energy environment of the singularities at the beginning and end of the universe. Thus, by the preceeding argument, we have that the number of universes N is on the order of 10^{43}. Hence, we have

$$a^2 = NMY10^{52} = 10^{209} \quad (4.13)$$

given that the number of particles within each universe is roughly 10^{80}, as cosmological observations for our universe suggest, and that each universe lasts for roughly 10^{10} years, on the order of current cyclical cosmological models, giving us the temporal velocity

$$a = 3 \times 10^{104} \quad (4.14)$$

C. Calculation of Temporal Barrier Strength

Given (4.13), we can now determine the strength of the temporal barriers separating alternate realities. Specifically, we have that the chance of the particle tunneling through such a temporal barrier, as in [4], is given by

$$exp\left(-2ma^2 L \sqrt{\frac{U}{E} - 1}\right) \quad (4.15)$$

where we assume that the temporal barriers, while tall are extremely thin, with a width in external time given by

$$L = \frac{\hbar}{|E|} \quad (4.16)$$

Thus, (4.15) reduces to

$$exp\left(-4\sqrt{\frac{U}{E} - 1}\right) \quad (4.17)$$

Now, we can consider the movements of the electron through the multiverse as analogous to a "random walk" on the real number line. A "bloc" is a period of internal time during which the electron remains confined to a single universe (between two specific barriers). The electron has an equal probability, assuming equal barrier strength, to tunnel to the external-time future or past, and as such the next bloc has an equal probability of being confined to the adjacent future universe, or past universe, of the previous bloc. Therefore, the relevant random-walk to this situation is that on the integers, with probability $p = 0.5$ of motion left or right. After N steps in this model, the farthest distance traversed from the starting point is proportional to $N^{1/2}$. Thus, if we desire the particle to populate 10^{43} universes, it must traverse 10^{86} blocs. There are therefore 10^{43} blocs per universe, meaning that each bloc, as per the above arguments, should include 10^{37} "laps" between the temporal barriers. Thus, we have

$$exp\left(-4\sqrt{\frac{U}{E}-1}\right) = 10^{-37} \quad (4.18)$$

meaning that

$$U = 2.3 \times 10^{211} \quad (4.19)$$

V. Universe-Dependent Temporal Flow

We have from (4.1) that the motion of the temporal wavepacket through external time, absent an imposed potential, is given by (in atomic units)

$$\phi(\tau,t) = \left(\frac{2c}{\pi}\right)^{1/4} \frac{exp\left(\frac{-c\tau^2 + i(l\tau - l^2 t/2ma^2)}{1 + 2ict/ma^2}\right)}{\sqrt{1 + 2ict/ma^2}} \quad (5.1)$$

Using the generalized Schrodinger equation (3.25), in conjunction with the corresponding temporal duality expressed in (1.14) (from the analogous separation process on the generalized Schrodinger equation), we have the temporal wavepacket

$$\phi(\tau,t) = \left(\frac{2c}{\pi}\right)^{1/4} \frac{exp\left(\frac{-c\tau^2 + i(l\tau - l^2 t/2mza^2)}{1 + 2ict/mza^2}\right)}{\sqrt{1 + 2ict/mza^2}} \quad (5.2)$$

where

$$z = \frac{|exp(i\mu)|_p}{exp(i\mu)} \quad (5.3)$$

with μ the parameter indexing different universes of the MDM multiverse, lying, as before, between π to 2π. Let us consider the speed of this wavepacket through external time. As before, we constrain the temporal momentum l of the above wavefunction such that the speed of the particle through external time, in our universe, is exactly 1. However, when complex path-discretization is invoked, as shown below, the temporal speed of the particle (not the temporal velocity, a constant) varies from universe to universe, the same initial temporal wavepacket (5.2 for $t = 0$) evolving differently for distinct versions of the general Schrodinger equation (3.25).

To ascertain the speed of the temporal wavepacket, we determine the maximum value of its probability density. This density is clearly proportional to

$$\left| exp\left(\frac{-c\tau^2 + i(l\tau - l^2 t/2mza^2)}{1 + 2ict/mza^2}\right) \right|^2$$

$$= exp\left(2Re\left(\frac{-c\tau^2 + i(l\tau - l^2 t/2mza^2)}{1 + 2ict/mza^2}\right)\right) \quad (5.4)$$

where the proportionality is given by the normalization function $A(t)$, dependent on the value of internal time. However, this factor $A(t)$, although adjusting the overall time-dependent proportionality of the wavefunction and the potential (3.12), will clearly not affect the position of the maximum of (5.2). This last, by (5.4), is given by the maximum of

$$Re\left(\frac{-c\tau^2 + i(l\tau - l^2 t/2mza^2)}{1 + 2ict/mza^2}\right) \quad (5.5)$$

This reduces to

$$-\frac{l^2t^2c\sin^2(u)}{m^2a^4b^2\left(\frac{4t^2c^2\cos^2(u)}{m^2a^4b^2}+\left(\frac{2tc\sin(u)}{ma^2b}+1\right)^2\right)}$$

$$-\frac{l^2t^2c\cos^2(u)}{m^2a^4b^2\left(\frac{4t^2c^2\cos^2(u)}{m^2a^4b^2}+\left(\frac{2tc\sin(u)}{ma^2b}+1\right)^2\right)}$$

$$-\frac{l^2t\sin(u)}{2ma^2b\left(\frac{4t^2c^2\cos^2(u)}{m^2a^4b^2}+\left(\frac{2tc\sin(u)}{ma^2b}+1\right)^2\right)}$$

$$+\frac{2ltc\tau\cos(u)}{ma^2b\left(\frac{4t^2c^2\cos^2(u)}{m^2a^4b^2}+\left(\frac{2tc\sin(u)}{ma^2b}+1\right)^2\right)}$$

$$-\frac{w\tau^2}{\left(\frac{4t^2c^2\cos^2(u)}{m^2a^4b^2}+\left(\frac{2tc\sin(u)}{ma^2b}+1\right)^2\right)}$$

$$-\frac{2tc^2\tau^2\sin(u)}{ma^2b\left(\frac{4t^2c^2\cos^2(u)}{m^2a^4b^2}+\left(\frac{2tc\sin(u)}{ma^2b}+1\right)^2\right)} \quad (5.6)$$

where

$$z = b\exp(ui) \quad (5.7)$$

The maximum of this expression in external time is given by

$$\tau = \frac{lt\cos(u)}{2tc\sin(u)+ma^2b} \quad (5.8)$$

We have that the value of $ma^2b \gg 2tc\sin(u)$, as for our single electron the value of the left side is roughly 10^{209}, while assuming a wavepacket of Planck-scale width, c is only roughly 10^{-54}. Thus, the velocity of the wavepacket is very nearly given as

$$\frac{l\cos(u)}{ma^2b} = \frac{Re(z)}{|z|} \quad (5.9)$$

Using the aforementioned value for l. We note that (5.8) reduces to the standard formula for the velocity, in terms of the momentum, when the path discretization is real and the dual spatial situation is considered.

For real discretization, (5.9) yields the unitless temporal speed of 1. However, when complex discretization is introduced, (5.9) leads to temporal speeds less than one. When the same initial wavefunction ((5.2) at $t = 0$) with the known value for l is allowed to evolve in some parallel universe, its movement through external time will generally be slower (relative to internal time) than in our own universe. Thus, each universe has a different rate of comparative temporal flow. Thus, allowing the particle to "quantum jump" to another universe would see its speed through time, compared to our own universe, decrease significantly by (5.9). Exploiting this property allows considerable potential applications.

VI. Superluminal Travel and Communication

The temporal speed factor (5.9) presents considerable ramifications on the nature and structure of this multiverse. Namely, a particle starting with the same initial temporal wavefunction will exhibit a lower temporal speed in the complex-discretized universes, compared to our own real-discretized universe. Thus, the single particle of the model progresses slower through external time in the alternate universes, compared to our own reality. This differential in temporal speed could pose considerable application, given a particular relationship between the relevant universes.

Namely, if the alternate realities of this combined theory exist "parallel" or "coexistent" in some sense, so that the future temporal barrier of one is smoothly topologically identified with the past barrier of the next, accessing these other realities might be possible. Such a scheme should assign the temporal potential barrier, being the border of two realities, the physical laws corresponding to one or the other. For example, the laws applying within each temporal barrier might correspond to the universe immediately preceeding it, in external time. The general principle, however, remains intact; we are engineering the external-time continuum so that successive universes in external time are parallel or coincident with each other, the temporal barrier linking external-time past and future serving as a way to topologically identify the future of one such universe with the past of the next. This programme follows the same rough principle as Roger Penrose's Cyclical Cosmology [1].

If accessing such parallel universes is a possibility, through a wormhole, space-time configuration, or some other contrivance, then the temporal factor of (5.9) could be used to achieve effective superluminal travel and communication. The continuum of external time, in accordance with the interpretation introduced in Section 1.B, would serve as the metric to determine the interaction points between such universes; namely, if an observer O leaves our universe U and enters a parallel universe U', at external time E in our universe and E' in U', and waits for an external time interval Y measured in U', then leaves, he will exit at external time E'+Y in U' and E+Y in U. That is, the total interval of external time traversed between entering and exiting the alternate universe is consistent with the frames of reference adopted in each separate universe. This is simply to say that, much as the most natural system would see spatial points in each separate universe "line up" in an analogous manner, so to do external time coordinates "line up," the that the total interval between entry and exit is the same in both realities. This is consistent with the general interpretation of external time serving as the metric for indexing the interaction between separate systems.

However, by (5.9) the same is not true of internal time. As the particle travels slower through external time in these other realities, the total interval of internal time experienced by the particle, between the external-time points corresponding to trans-universal entry and exit, is greater than the corresponding external-time in our universe. Such a journey between universes is consistent with the one-electron model we have been using, the movement of an apparatus from one universe to another simply being the corresponding movement of each of its constituent particles, each being distinct internal-time states of the one particle. As internal time determines the evolution of the particle's state, by (1.14), a starship transported into a parallel reality could accelerate to approach the speed of light, the latter determined by the ratio between spatial distance and internal time. However, due to the temporal differential of (5.9), the effective spatial velocity relative to external time could be very much greater, proportional to the factor $\frac{Re(z)}{|z|}$. Hence, effective superluminal travel is possible in this model.

The notion of "hyperspace" is a trope well-explored in the context of science-fiction, where superluminal travel is possible in some alternate or parallel dimension. Generally, this dimension, "hyperspace," is constituted so that small spatial distances in hyperspace correspond to comparatively large distances in "real space." Thus, the starship enters hyperspace, traverses a relatively short distance, and exits hyperspace at the point corresponding to its destination in "real space," thus completing a journey in effectively superluminal time [3]. The present model is similar, with these complex-discretized dimensions, "phasespace," offering conduits for superluminal travel. Here, however, the mechanism is different; rather than spatial distance being the relevant factor, temporal intervals are. The ship enters phasespace, makes a journey to a distant star at, say 50% light speed, taking 100 years in phasespace; when the ship re-enters "real space," however, only 10 years will have passed, allowing the ship to effectively travel at five times the speed of light. Use of constant acceleration might take advantage of time dilation, so that even the time measured within the ship might be comparatively small to that measured by the stationary observer in phasespace; however, consideration of this latter point would require combining quantum temporal dynamics and complex-discretization with relativistic quantum mechanics, which has not, as yet, been done.

VII. Conclusion

Work completed in [4] was developed from the path-integral framework, rather than assuming a generalized Schrodinger equation ad hoc. The one-electron model of quantum temporal dynamics, where a two-time theory provides a quantum model for time's passage and the reflections of a single particle between two barriers produce all the particles seen in the universe, was combined with the multiphase discretization metaverse, where complex-discretization of functional integrals produces a panoply of parallel universes. The temporal cycles emerging from quantum temporal dynamics were postulated to correspond to the alternate realities of MDM. By using the new path-integral framework from QTD, complex-discretization was built into MDM, offering the laws of physics applying to these separate realities. Some conclusions were drawn from these results, including the likely sizes of certain constants of the theory, the sizes of the temporal barriers, and the differential in temporal flow between universes.

As a result of this last property, we discovered that superluminal travel would be possible if these alternate realities could be accessed. Given that a certain topological identification exists so that the temporal cycles are coincident in external time, so that these universes can indeed be accessed, travel through an alternate dimension could proceed at a speed greater than that of light. This theory, however, is nonrelativistic, and the development of a complex-discretized quantum field theory, as well as a relativistic version of QTD, could allow these results to be extended significantly.

References

[1] Araujo, Jennen, Pereira, Sampson, Savi. "On the Spacetime Connecting Two Aeons in Conformal Cyclic Cosmology." 2015.

[2] Feynman, Richard. <u>Quantum Mechanics and Path Integrals</u>. New York: McGraw-Hill, 1965.

[3] Varieschi, Gabrielle and Zily Burstein. "Conformal Gravity and the Alcubierre Warp Drive Metric." 2012.

[4] von Abele, Julian. "A One-Electron Theory of Nonrelativistic Quantum Temporal Dynamics ." 2017.

[5] —. "Generalization of Path-Integration to a Complex Time-Discretization Index ." 2016.

[6] Srednicki, Mark. "Quantum Field Theory." University of Santa Barbara. 2006.

Superluminal Communication and Travel in the Quantum Temporal Dynamics Multiverse

Abstract:

In the present paper, we extend previous results developed in [4], concerning the multiverse of quantum temporal dynamics. In that paper, we combined the formulation of quantum temporal dynamics with the multiverse framework arising from complex path-integral discretization, resulting in a cyclical cosmology with different cycles corresponding to alternate universes with distinct laws of physics. There, we discovered that the rate of temporal flow (the "temporal speed") depends upon cosmological cycle, and hence that access to other cosmological cycles could permit faster-than-light travel. In this paper, we expand upon the superluminal implications of this theory, determining the effective superluminal speed with which signals can propagate when these dimensions can be accessed. Furthermore, the quantum behavior of these alternate realities could allow rapid subluminal travel within each universe, thus permitting effectively superluminal travel with respect to our universe.

I. Introduction

A. Overview of the Temporal Multiverse

In [4] we considered a cyclical cosmology, comprised of a sequence of parallel universes that exist at alternate quantum phase. This theory was a result of combining multiphase discretization, as developed in [5], where the time-discretization index of the quantum functional integral is made complex, with quantum temporal dynamics, as in [3], where quantum mechanics is applied to the passage of time, and a dichotomy between time internal and external to a system is proposed. In [5], we took the standard formula for the path-integral in quantum theory,

$$P(a,b) \cong \lim_{n \to \infty} \iiint \ldots \int exp\left(\frac{i}{\hbar}S\right) dx_1 \ldots dx_{n-1} \quad (1.1)$$

and allowed $n \in \mathbb{N} \to z \in \mathbb{C}$, yielding a spectrum of different propagators depending on how z approaches infinity in the complex plane. We proposed that these propagators correspond to distinct universes. This multiverse lent itself naturally to the quantum temporal dynamics developed in [3], where we applied quantum mechanics to the passage of time, promoting time from parameter to observable. This framework necessitated the distinction between the time *internal* to a system, indexing the change of its quantum state, and *external* to a system, indexing the relationship between systems. Applying this temporal dichotomy, we developed a possible framework for the one-electron universe, where the movement of a single particle forward and backward through time produces all the particles in the universe. Thus, the potential at any point in space-time is due entirely to the influence of this single particle. The formula of the potential in this theory is given as

$$V(x,\tau) = R(\tau) + \iiint |\psi(x',\tau';t')|^2 V(x - x')\delta_\varepsilon(\tau - \tau') dx' d\tau' dt' \quad (1.2)$$

Where x is the spatial co-ordinate, τ the external time variable, and t the internal time parameter. Here, V determines the standard electrostatic potential over the spatial distance, while δ_ε is a Gaussian of width ε. ψ is a "temporal wavefunction" that determines the probability density of the particle existing, not only in a certain spatial range, but also in a range of external time. $R(\tau)$ are large temporal barriers that reflect the particle, resulting in the appearance of many particles at any given value of external time.

Using a path-integral formulation of temporal dynamics, where external time is treated as a rescaled spatial dimension, as in [4], we developed a Schrodinger equation for the evolution of the temporal state,

$$-\frac{\hbar^2}{2m} \Box_a \psi(x,\tau;t)$$
$$+ \Big(R(\tau)$$
$$+ \iiint |\psi(x',\tau';t')|^2 \, V(x-x') \delta_\varepsilon(\tau - \tau') \, dx' d\tau' dt' \Big) \psi(x,\tau;t)$$
$$= \exp(-i\mu) \, i\hbar \frac{\partial \psi(x,\tau;t)}{\partial t} \quad (1.3)$$

Where $\Box_a = \nabla^2 - \frac{1}{a^2}\frac{\partial^2}{\partial t^2}$, a being a constant we dubbed the "temporal velocity." As such, the evolution of the wavefunction at any given internal time t depends in principle on its value at all other times t', a notion certainly consistent with our consideration of a one-electron universe, where all particles are instantiations of a single particle at distinct internal times. (1.3) readily separates as

$$-\frac{\hbar^2}{2m} \nabla^2 \theta(\mathbf{r},t) + V(\mathbf{r}) \theta(\mathbf{r},t) = i\hbar \frac{\partial}{\partial t} \theta(\mathbf{r},t) \quad (1.4)$$

$$\frac{\hbar^2}{2ma^2} \frac{\partial^2}{\partial \tau^2} \phi(\tau,t) + R(\tau) \phi(\tau,t) = i\hbar \frac{\partial}{\partial t} \phi(\tau,t) \quad (1.5)$$

Where
$$\psi(\mathbf{r},t;T) = \theta(\mathbf{r},T) \phi(t,T) \quad (1.6)$$

It is natural to consider the existence of space-time beyond the aforementioned temporal barriers. An intriguing possibility suggested both in [3] and in [4] is that alternate universes exist in the internal-time future and past, marked by a succession of other potential barriers $R(\tau)$. Between any two of these barriers lies another reality. By quantum tunneling through a barrier, a particle can reach an adjacent universe. As this single particle reflects through this potential labyrinth, it systematically fills all these universes. This multiverse framework offers an intriguing system for effecting the multiverse of complex-phase discretization as in (1.1). Namely, each different universe occupying an internal-time range in this cyclical cosmology might correspond to an alternate complex phase of the discretization index.

In [4], we applied (1.1) and the framework developed in [5] to quantum temporal dynamics, resulting in the generalized Schrodinger equation

$$-\frac{\hbar^2}{2m|\exp(i\mu)|_p} \Box_a \psi(x,\tau;t) + \{|\cos(\mu)|^p$$

$$- |\sin(\mu)|^p\} (|\cos(\mu)|^p + |\sin(\mu)|^p)^{\frac{1}{p}-1} \Bigg(R(\tau)$$

$$+ \iiint \frac{|\psi(x',\tau';t')|^2}{\int_{x'-\tau'} |\psi(x',\tau';t')|^2 dx' d\tau'} V(x-x') \delta_\varepsilon(\tau$$

$$- \tau') \, dx' d\tau' dt' \Bigg) \psi(x,\tau;t) = \exp(-i\mu)\, i\hbar \frac{\partial \psi(x,\tau;t)}{\partial t} \quad (1.7)$$

Where μ is a parameter indexing the phase of the discretization index and p determines the norm used in the generalization of the functional integrand. Using an anthropic argument, concerning the structure of nucleons and the range of quantum uncertainty in quark position within the nucleon, we determined from (1.7) that the value of the temporal velocity is roughly 3×10^{104} in atomic units, that the size of the temporal barriers is roughly 2.3×10^{211}, and that the number of universes in the cosmological system is on the order of 10^{43}.

B. Superluminal Implications

An interesting property of this multiverse, as we discovered in [4], is that the rate of temporal flow, the "speed" of the particle through external time, varies from universe to universe. That is, if a quantum temporal state ψ is transported, instantaneously, to another reality, its evolution in internal time will exhibit a speed through external time generally less than 1. This speed is determined by the evolution of the temporal factor $\phi(\tau, t)$ in the relevant universe. In general, this "temporal speed" is proportional to

$$\frac{Re(z)}{|z|} \quad (1.8)$$

with z being the complex mass coefficient $\frac{|\exp(i\mu)|_p}{exp(i\mu)}$. As such, the temporal speed in other realities is generally less than in our own, opening up the possibility of using these dimensions for superluminal travel.

For, if these universes are topologically identified so that the external-time future of one is linked to the past of the next, and the system "wraps around" so that successive universes co-exist, as proposed in [4], then travel to these other realities might be a possibility. A starship could simply enter "phasespace," a parallel universe at an alternate quantum phase, travel with subluminal velocity through phasespace, and exit at the point corresponding to its destination in realspace. Measured relative to the ship, an internal time t will have passed (not the relativistic proper time, but in this non-relativistic model the time measured by all particles in that universe), and thus in accordance with the aforementioned result an external time $\frac{Re(z)}{|z|}t$ will have passed. The internal time determines the evolution of the particle's quantum state, in accordance with (1.4), and thus the distance the ship can travel, being composed of many instantiations of the particle at distinct internal times. However, external time indexes the interaction between the ship and external systems, and hence once the ship injects into "realspace," a smaller length of time will have passed for the realspace observer compared to the phasespace observer. Thus, for example, the ship could travel close to the speed of light in phasespace, travel for a hundred years, and drop out of phasespace at its destination in realspace where only a year has passed, thus effectively travelling at nearly 100c.

In addition to allowing such superluminal travel, phasespace also permits violations of the conservation of momentum, and therefore an apparatus for approaching high subliminal speeds with little energy cost. Although this model is non-relativistic, a rough consideration of special relativity would seem to suggest that the proper time relative to the ship is significantly lower than the internal time of the phasespace observer, thus meaning that both the realspace observer, and the one on the ship, experience comparatively small durations of time during the superluminal journey. In the succeeding section we consider how phasespace permits violations of momentum conservation, and the impact on intra-universe propulsion. In the following sections, we develop the specifics of the mechanics of phasespace travel.

II. Momentum Generation in Parallel Universes

A. Time Reversability

Given an initial quantum wavefunction

$$\sum \psi_i \quad (2.1)$$

in a system with energy eigenvalues $E_1, E_2 \ldots$, the time-evolution is given by

$$\sum \psi_i \exp\left(-\frac{i}{\hbar} E t\right) \quad (2.2)$$

and we have that

$$\sum \psi_i \left(\exp\left(-\frac{i}{\hbar} E t\right)\right)^* = \sum (\psi_i)^* \exp\left(-\frac{i}{\hbar} E t\right) \quad (2.3)$$

Thus, this implies

$$\sum \psi_i \exp\left(\frac{i}{\hbar} E t\right) = \sum (\psi_i)^* \exp\left(-\frac{i}{\hbar} E t\right) \quad (2.4)$$

meaning that, in standard non-relativistic quantum mechanics, systems evolve in reversed time in the same fashion as systems evolving forward in time, but with momenta reversed. Thus, the behavior of the particle during a leg of its journey with negative temporal speed is exactly the same as those with positive temporal speed, as measured from an observer at a certain point in external time. However, this is only true in a universe with real discretization index; in a parallel universe, the relation (2.3) will generally not reduce to (2.4), as the internal-time evolution factor is given by

$$T(t) = \exp\left(-\frac{i}{\hbar} E \exp(i\mu) t\right) \quad (2.5)$$

in this case. This is easily demonstrated by developing an analogous, complex-discretized equation for (1.4) from separating (1.7), and applying the procedure explicated in [5]. Generally, (2.5) will depress a particle's expected momentum as it evolves in internal time.

However, if the particle is moving with negative temporal speed, its expected momentum will be perceived to grow exponentially, in accordance with

$$\exp\left(\frac{i}{\hbar} E \exp(i\mu) \tau/v\right) \quad (2.6)$$

where v is the magnitude of the temporal speed, relative to external time.

B. Generating Momentum in a Parallel Universe

Let us consider, for the succeeding argument, a free particle where the potential is set to zero. Suppose that its spatial wavefunction is given by

$$\theta(\mathbf{r},t) = \int \sigma(E) \psi_E \, dE \quad (2.7)$$

where ψ_E represents the eigenvector state at energy E. By the above, its spatial quantum state will evolve relative to external time as

$$\theta(\mathbf{r},t) = \int \sigma(E) \psi_E \exp\left(\frac{i}{\hbar} E \exp(i\mu) \tau/v\right) dE \quad (2.8)$$

As described in [4], the actual probability density is only proportional to the wavefunction, the constant of proportionality changing with internal time so as to maintain normalization. By (2.7), however, the energy distribution $\sigma(E)$ of the quantum state will change proportionately with the factor $\exp\left(\frac{i}{\hbar} E \exp(i\mu) \tau/v\right)$ as external time moves forward. Thus, for instantiations of the particle in this universe with reverse temporal orientation, the expected momentum will generally shift right, in violation of momentum conservation which holds in our universe. This property could be utilized by a starship to generate momentum at low energy cost, therefore permitting fast subluminal travel (relative to internal time) within the universe.

The starship is composed of numerous instantiations of the single particle, corresponding to distinct ranges of internal-time. As such, the internal time t corresponding to the evolution of this system's state should be directly proportional to the external time τ, but rescaled by the magnitude of the temporal speed as in τ/v. That is, each constituent particle state evolves through external time with the same magnitude of temporal speed v, but different temporal directions; as such, time-reversed particles are perceived by the shipbound observer to move backward in time with a speed of 1.

Thus, we have the effective energy proportionality factor

$$\exp\left(\frac{i}{\hbar} E \exp(i\mu) t\right) \quad (2.9)$$

Let us consider the behavior of a hydrogen atom with reversed temporal orientation. In atomic units, we have (2.9) reduce to

$$\exp(1000 \, s^2 t) \quad (2.10)$$

using an order-of-magnitude approximation for the exponent, where s is the atom's spatial velocity. Let us take an initial velocity distribution

$$\sigma = \exp(-(10^5(s - 10^{-6}))^4) \quad (2.11)$$

Corresponding to a particle whose expected velocity ranges about 10 m/s around the origin, with a maximum of the distribution slightly displaced from zero at around 1 m/s. The evolution of this distribution is almost exactly the same as that of $\exp(-(10^5 s)^4))$, except that rather than two maxima, only the rightward maximum of the evolved wavefunction is of any actual relevance. As such, we can approximate the evolution of (2.11) by examining the average velocity of

$$\exp(-(10^5 s)^4)) \exp(1000 \, s^2 t) \quad (2.12)$$

which is roughly $s = \frac{\sqrt{t}}{500000000}$, given by the maximum of the evolved distribution. Hence, integrating, we determine that the spatial velocity of the particle after 1 meter is roughly 7,000 m/s. Depending on the simplifying assumptions made in the foregoing calculation, and the quantum phase of the corresponding universe, the true value could be anywhere from 1-10,000 m/s. Regardless, the point is that atoms in reverse temporal orientation can be assembled in such a quantum state that they anomalously increase in momentum without an input of energy. Such a system could be used by a starship in "phasespace" to accelerate close to the speed of light within the parallel universe.

The superluminal applications of entering alternate universes arise due to the divergent temporal speed factor as expressed in (1.8); namely, that a quantum state injected into a parallel universe will generally evolve slower through external time than in our own universe. The unique properties of complex-discretized quantum theory, however, unveil further applications. By taking advantage of the violation of momentum conservation in these parallel realities, a hypothetical starship could accelerate atoms in reverse temporal orientation by preparing them in a certain quantum state as in (2.11), and simply using their natural state evolution. These particles could thus be used as a source of momentum to drive the starship forward without fuel. Not only does phasespace allow faster-than-light travel relative to realspace, but within phasespace, a ship can easily accelerate close to the speed of light.

III. Using Alternate Dimensions for Superluminal Travel

In the cosmological model that we are envisioning, universes exist in discrete succession, following each other along the external time continuum. We imagine that the universes are topologically identified with their predecessors and successors so that they "wrap around" in external time, superimposed over the same range of external time. Thus, it might be possible for a ship to "jump" from one multiverse "layer" to another. Suppose a starship is capable of traversing N such discrete dimensional steps. Thus, it can jump through N universes when the drive mechanism is activated. It makes a subluminal journey relative to internal time in its destination universe (which, as we explained before, corresponds not to relativistic proper time, but to a stationary observer placed in that universe), and jumps back to our universe at the point corresponding to its ultimate destination, in such a fashion that very little time will have passed in our universe. Its effective superluminal speed is dependent upon the temporal scaling factor (1.8) corresponding to the relevant universe.

Let us suppose that the parameter μ has its value randomly and uniformly distributed over the constituent universes of the multiverse (in accordance with [5] its value falls between π and 2π). The parameter p could have its value distributed in accordance with the geometrical distance in Fig. 1 of [5] between the boundary curves of the multiverse. Given the "dimensional capacity" N of our starship, it can be expected by the foregoing assumptions that the universe with the highest temporal scaling factor, that it can reach, has a value of μ of roughly

$$\mu \in \left(\frac{3\pi}{2} - \frac{\pi}{2N}, \frac{3\pi}{2} + \frac{\pi}{2N}\right) \quad (3.1)$$

Thus, the expected temporal scaling factor the ship can reach is approximately

$$\frac{Re(z)}{|z|} = \frac{Re(\frac{|\exp(i\mu)|_p}{\exp(i\mu)})}{\left|\frac{|\exp(i\mu)|_p}{\exp(i\mu)}\right|} \approx \cos\left(\frac{3\pi}{2} + \frac{\pi}{2N}\right) = \cos\left(\frac{\pi}{2} - \frac{\pi}{2N}\right) \quad (3.2)$$

The central principle here being that, as the values of the μ parameter are randomly distributed from universe to universe, access to more universes makes it more likely the ship can take advantage of a layer of the multiverse with very high temporal scaling factor. Indeed, $\mu = \frac{\pi}{2}$ is associated with infinite (or zero) scaling factor, and the closer the ship can approach this value, the greater will be its effective speed. In accordance with (3.2), this effective speed is

$$\frac{1}{\cos\left(\frac{\pi}{2} - \frac{\pi}{2N}\right)} \quad (3.3)$$

times the speed of light.

In phasespace, as (2.9) makes clear, the standard momentum conservation is violated, and particles see their momenta anomalousy increase or decrease, relative to some external system, depending on their temporal orientation (whether their temporal speed is positive or negative). Indeed, this change in the energy profile of quantum states over internal time was used in [4] to provide limits on the value of the μ parameter in our own universe, and therefore the probable number of layers of the multiverse. As such, it is clear that normal matter will not retain its structure in phasespace, and that a ship must be protected to survive in this environment with alternate laws of physics. Projecting a bubble of normal space-time around the ship, much like the "warp bubble" used in the Alcubierre drive, could allow the ship to survive in phasespace [1]. During transit, this space-time field would be maintained and projected aroud the ship, creating a bubble of normal space where standard physical laws apply. Failure of this bubble would mean the collapse of the ship's structure at a fundamental level.

To propel itself through phasespace, a ship would rely on "momentum generators" using the same principle as that introduced in the previous Section; namely, that a particle prepared in a certain quantum state, and released into phasespace, will see its momentum increase without any energy input (provided that it exists at a reverse temporal orientation). A mechanism could prepare hydrogen atoms from a fuel store at such reverse orientation into a Bose-Einstein condensate with an initial wavefunction like (2.11), and inject them into a "field cavity" where the space-time bubble has not penetrated. In accordance with the above calculation, such a hydrogen atom, when prepared with an initial speed of roughly 10 m/s, would accelerate to roughly 5,000 m/s after 1 meter in the cavity. Once the atoms leave the cavity, a membrane or other device could be used to reflect them and extract momentum for the ship.

Some recent work suggests that wormholes might bridge universes in certain multiverse scenarios [2]. Indeed, a starship might use a wormhole or other space-time configuration as a means to reach these other realities. Injecting into phasespace would likely carry a great energy cost, as would re-emerging in normalspace. A ship would need equipment to create a gateway to phasespace, as well as technology to project a realspace bubble or field for transit, in addition to momentum generators for propulsion within phasespace itself. While these hypothetical considerations of faste-than-light travel are certainly speculative, it is intriguing that superluminal travel seems perfectly consistent with a compelling theoretical framework.

References

[1] McMonigal, Brendan, Geraint Lewis and Phillip O'Byrne. "The Alcubierre Warp Drive: On the Matter of Matter." 2012.

[2] Novikov, I.D., A.A. Shatskiy and D.I. Novikov. "The Wormholes and the Multiverse." 2014.

[3] von Abele, Julian. "A One-Electron Theory of Nonrelativistic Quantum Temporal Dynamics." 2017.

[4] —. "A Path-Integral Formulation of Nonrelativistic Quantum Temporal Dynamics, and Implications on the Multiphase Discretization Metaverse Model ." 2017.

[5] —. "Generalization of Path-Integration to a Complex Time-Discretization Index." 2016.

Made in the USA
Middletown, DE
17 June 2022